Plant Cell Culture

Plant Cell Culture

Managing Editor: A. Fiechter

with 51 Figures and 18 Tables

Akademie-Verlag · Berlin 1985

Die Originalausgabe erscheint im Springer-Verlag
Berlin—Heidelberg—New York—Tokyo
als Volume 31 der Schriftenreihe
Advances in Biochemical Engineering/Biotechnology

Vertrieb ausschließlich für die DDR und die sozialistischen Länder

Erschienen im Akademie-Verlag · Berlin, DDR-1086 Berlin, Leipziger Straße 3—4
Lizenznummer: 202 · 100/545/85
Printed in the German Democratic Republic
Gesamtherstellung: VEB Druckerei „Thomas Müntzer", 5820 Bad Langensalza
Umschlaggestaltung: Karl Salzbrunn
LSV 1315
Bestellnummer: 763 563 6 (3070/31)

09000

Table of Contents

Biosynthesis of Secondary Products by Cell Cultures of Higher Plants

L. A. Anderson, J. D. Phillipson, and M. F. Roberts
Department of Pharmacognosy, The School of Pharmacy, University of London, 29–39 Brunswick Square, London, WC1N 1AX, UK

Biosynthetic studies of alkaloids, phenols and terpenes utilizing plant cell culture techniques have been reviewed from literature dating mainly from 1980. Many of the research papers deal with the investigations into the alkaloids of *Catharanthus roseus* but there are a number of significant publications in other areas of alkaloid biosynthesis, viz. other indole alkaloids, isoquinolines and quinolizidines. Flavonoid biosynthesis continues to be an active area of research in which much use has been made of plant cell cultures. Individual steps in secondary product biosynthesis tend to be the focus of attention with particular emphasis on the isolation and characterization of the enzymes involved. The use of plant cell culture for biosynthetic studies has been greatly enhanced by the development of sensitive analytical techniques.

1 Introduction

Although plant cell cultures do not necessarily produce the same secondary metabolites as their parent plants, they are nevertheless valuable for the investigation of biosynthetic problems [1–5]. Cell suspensions are used in preference to callus cultures in such studies because of the easier administration of precursors and the extraction of products. In some aspecıs, plant cell cultures possess advantages over intact plants for biosynthetic studies, e. g.,

a) Cultures can be grown under standard conditions for short growth cycles and are not subject to seasonal variation.
b) Cultures are less complex in organization than the entire plant and hence permeability, translocation and segregation of precursors and products do not present the problem of incorporation which are sometimes encountered in whole plants.
c) Purified enzymes and active cell-free systems can be prepared more easily from cell cultures.

The obvious disadvantages of plant cell cultures for biosynthetic studies include the necessity for working under aseptic conditions and the fact that such cultures may not produce the same secondary metabolites as the parent plant. The synthesis of some secondary metabolites is thought to be connected with cell differentiation or with the organization of tissues so that it is to be anticipated in biosynthetic studies that particular secondary metabolites may not be produced in cultures [1,6].

The major routes of biosynthesis for many secondary metabolites are now largely understood but what is required is the complete elucidation of pathways by determining each intermediate and the characterization of the enzyme(s) involved in its formation.

The combination of cell suspension cultures and highly specific assay procedures such as radioimmunoassay have proved invaluable and this has been well demonstrated in such areas as indole alkaloid biosynthesis [7]. The biosynthesis of polyketides, phenolics, terpenes and alkaloids has been reviewed in 1983 based on literature published during the period 1979 to 1981 and covers some references to plant cell culture techniques [8]. Furthermore, the previous literature devoted solely to the use of plant cell cultures for the study of biosynthetic pathways in higher plants has been reviewed extensively up to 1979 [1-5] and consequently literature references mainly from 1980 have been cited in this text.

2 Alkaloids

2.1 Indoles

Heterotrophic, photomixotrophic and photoautotrophic cell suspension cultures of *Peganum harmala* have been analyzed for harman-type alkaloids [9]. Only heterotrophic cultures produced these alkaloids and the presence and yield of individual alkaloids was influenced by the plant hormones added to the media. Evidence has been presented for the biosynthetic sequence of tryptophan → tryptamine → serotonin taking place in cell suspension cultures of *P. harmala* [10]. From feeding experiments and enzymatic measurements, it has been concluded that decarboxylase activity is a regulatory control for β-carboline alkaloid biosynthesis. [^{14}C] Labelled gramine has been fed to barley *(Hordeum vulgare)* cell suspension cultures and a series of products identified. The degradation sequence commences with demethylation to yield methylaminomethylindole and aminomethylindole and is followed by oxidative deamination to yield indole-3-aldehyde which is then either oxidized or reduced to form the corresponding acid or alcohol [11]. The majority of biosynthetic studies of indole alkaloids utilizing plant cell cultures have been concerned with the iridoid alkaloids and have concentrated on *Catharanthus roseus*. The clinically useful antitumor dimeric indole alkaloids vinblastine and vincristine are expensive to produce from whole plants and their potential production by plant cell cultures is an attractive possible alternative source. The first committed step in the biosynthesis of these alkaloids is the coupling of tryptamine (*1*) and the monoterpene secologanin (*2*) to yield the gluco-alkaloid, strictosidine (*3*) (Fig. 1). The enzyme responsible for this process in strictosidine synthase [7] which has been immobilized on CNBr-activated sepharose [12]. It is now possible to synthesize gram quantities of strictosidine for biosynthetic studies whereas previously this compound has proved difficult to prepare and to purify.

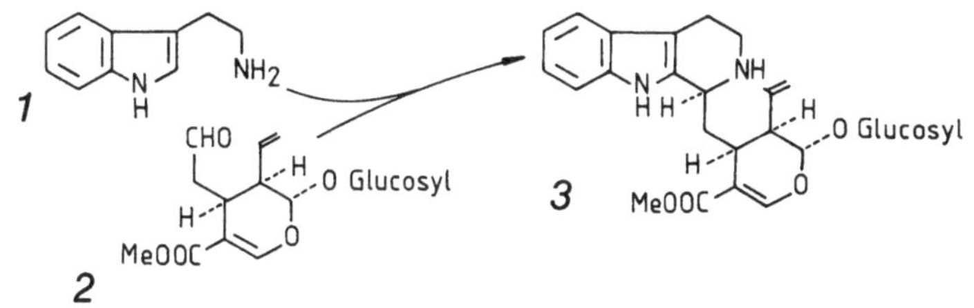

Fig. 1. The formation of strictosidine (3), the key intermediate in iridoid-indole alkaloids, from tryptamine (1) and secologanin (2)

One of the problems encountered in studying biosynthetic pathways is the low incorporation of labelled precursors into intact plants. This problem has been overcome for monoterpene incorporation into indole alkaloids by the use of cell suspension cultures of *C. roseus* [13]. Deuterated 10-hydroxygeraniol and 10-hydroxynerol have been incorporated into ajmalicine and strictosidine lactam in approximately 50% and 80% yields, respectively. Based on further experiments, it has been concluded that 9,10-dihydroxygeraniol, 9-oxo-10-hydroxygeraniol and 9,10-dioxogeranial are intermediates in the biosynthesis of ajmalicine. These experiments have led to a proposed biosynthetic pathway between geraniol and loganin [14]. A monoterpene hydroxylase has been isolated from cell suspensions of *C. roseus* and is unlike its counterpart in seedlings because a significant portion of the enzyme appears to be either soluble or only loosely bound [15].

An increasing number of monomeric monoterpenoid alkaloids continue to be isolated from cell suspension cultures of *C. roseus*. In one particular study, 12 alkaloids were isolated from *C. roseus* cell suspensions and 14 alkaloids from *C. ovalis* suspensions [16]. In all, eighteen alkaloids were reported from these two species and they represented the major types of *Corynanthe*, *Strychnos*, *Aspidosperma* and *Iboga* alkaloids. The ability of *C. roseus* cell suspensions to produce such a range of monomeric indole alkaloids has been amply demonstrated by several laboratories [17–23, 26–30]. These studies indicate that the monomeric alkaloids can be produced readily whereas the dimeric alkaloids are not so easily formed. [3-^{14}C]-

DL-tryptophan has been incorporated into akuammicine, catharanthine and vindoline by suspension cultures of *C. roseus* [17] and a number of distinct cell lines have been produced [18–20]. One particular strain of *C. roseus* culture has yielded an alkaloidal extract with antimitotic activity but no dimeric alkaloids were detected [22]. In one detailed investigation [23] a particular cell line of *C. roseus* (PRL 200) has been subjected to time-course studies, for alkaloidal yield and constituents, over a 27 day period, utilizing different media conditions. This cell line accumulates catharanthine in particular, together with 9 other indole alkaloids including strictosidine lactam. This latter alkaloid is usually regarded as an artifact produced readily from strictosidine (e.g., in vitro at pH 7.5) however, because of its high yield in this cell line it is possible that it may be formed enzymatically [23].

Cell-free systems from *C. roseus* leaves have been demonstrated to incorporate [2-^{14}C] tryptamine (*1*) and secologanin (*2*) into vindoline (*4*) [24]. The same cell-free system was utilized to couple vindoline (*4*) and catharanthine (*5*) to yield the dimeric 3′,4′-anhydrovinblastine (*6*) which in turn was converted to the natural dimeric alkaloids leurosine, catharine and vinblastine (*7*) (Fig. 2). These results corroborate an independent and simultaneous study in which it was shown that 3′,4′-anhydrovinblastine is converted to vinblastine by cell-free preparations of *C. roseus* prepared from whole plants [25] and from cell suspension cultures [26]. The successful de novo production of these dimeric alkaloids by plant cell culture techniques has proved to be difficult to achieve.

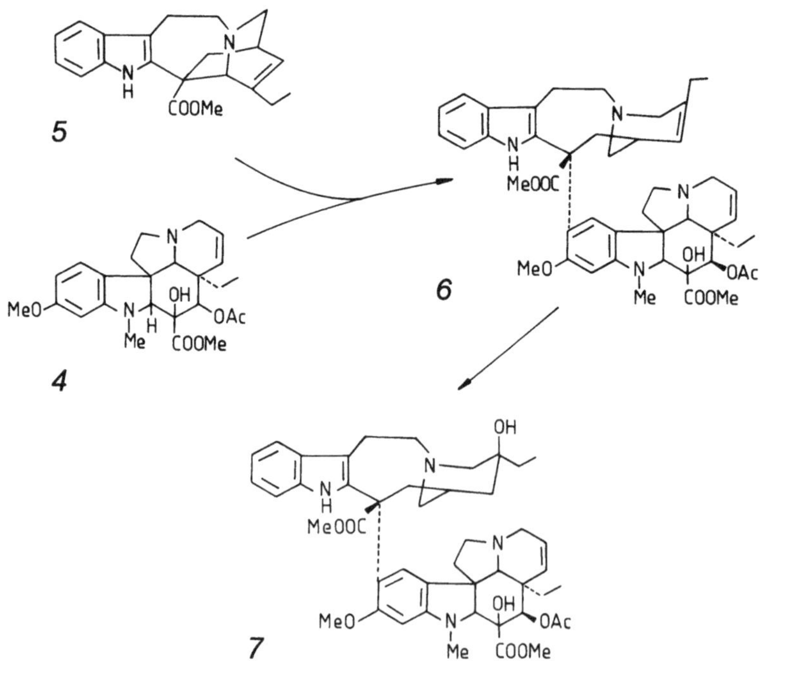

Fig. 2. The conversion of vindoline (4) and catharanthine (5) into 3′,4′-anhydrovinblastine (6) and vinblastine (7) by cell-free systems of *Catharanthus roseus* leaves

A cell line of *C. roseus* suspension cultures (PRL 953) which does not produce vindoline, failed to convert vindoline (*4*) and catharanthine (*5*) into dimeric alkaloids [27]. Yet another cell line of *C. roseus* ("916") which lacks the ability to produce the characteristic alkaloids of the plant, has been shown to possess the enzymes which are capable of converting 3′,4′-anhydrovinblastine (*6*) into leurosine and catharine [28–30].

Although it has proved difficult to produce dimeric indole alkaloids by *C. roseus* cultures, two novel dimeric alkaloids, voafrine A (*8*) and voafrine B (*9*), have been isolated from cell suspension cultures of *Voacanga africana* [31] (Fig. 3). It is proposed that these alkaloids are produced enzymatically in the cultures and are not artifacts produced because of the instability of tabersonine, the monomeric component of these alkaloids.

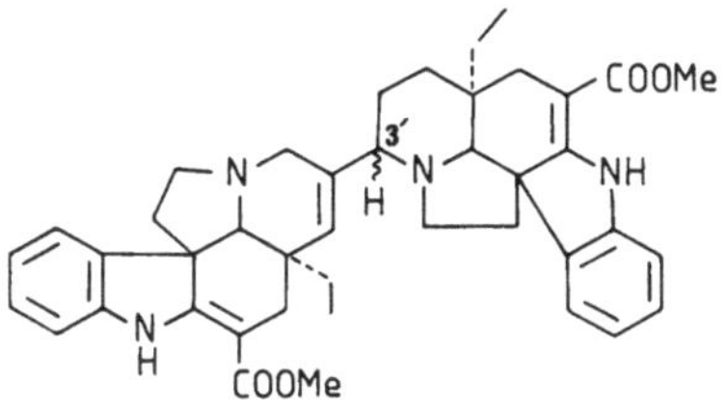

8 C-3′H α configuration
9 C-3′H β configuration

Fig. 3. Dimeric indole alkaloids isolated from cell suspensions of *Voacanga africana*

Earlier studies on the alkaloids of *C. roseus* cell suspensions showed that ajmalicine and other heteroyohimbine alkaloids (*Corynanthe*-type) are readily produced. The results of tracer feeding experiments led to the assumption that geissoschizine (*10*) was a central component in the biosynthesis of all of the major types of alkaloids found in *C. roseus*. The use of cell-free systems prepared from tissue cultures has given results which indicate that geissoschizine is only indirectly involved in the biosynthesis of heteroyohimbine alkaloids. Geissoschizine dehydrogenase catalyses the dehydrogenation of geissoschizine (*10*) at the C-21 α position to yield 4,21-dehydrogeissoschizine (*11*) which is a central intermediate in the biosynthesis of heteroyohimbine alkaloids. This enzyme which has now been isolated from a suspension culture of *C. roseus*, has been partially purified and characterized [32]. 4,21-Dehydrogeissoschizine (*11*) is converted to cathenamine (*12*) (19(R),19(S)-isomers and iminium forms) (*13*) which are the direct precursors of ajmalicine (*14*), 19-epi-ajmalicine (*15*) and tetrahydroalstonine (*16*) (Fig. 4). Previously, it had been shown that 17-hydroxy-19-epi-cathenamine, which is known to be a precursor of 19-epi-ajmalicine (*15*) under biomimetic conditions, was not involved in its biosynthesis by cell-free extracts [33]. The enzymatic synthesis of ajmalicine (*14*) and related heteroyohimbine alkaloids by means of cell cultures and cell-free systems has been lucidly reviewed [34]. Subsequently, crude enzyme extracts obtained from *C. roseus* cell suspension cultures have been used to follow the steric course of hydrogen transfer during the formation of the 3α-heteroyohimbine alkaloids [35].

The biosynthetic studies undertaken on cell suspensions and cell-free extracts of *C. roseus* have been facilitated by sensitive analytical techniques. Preparation of

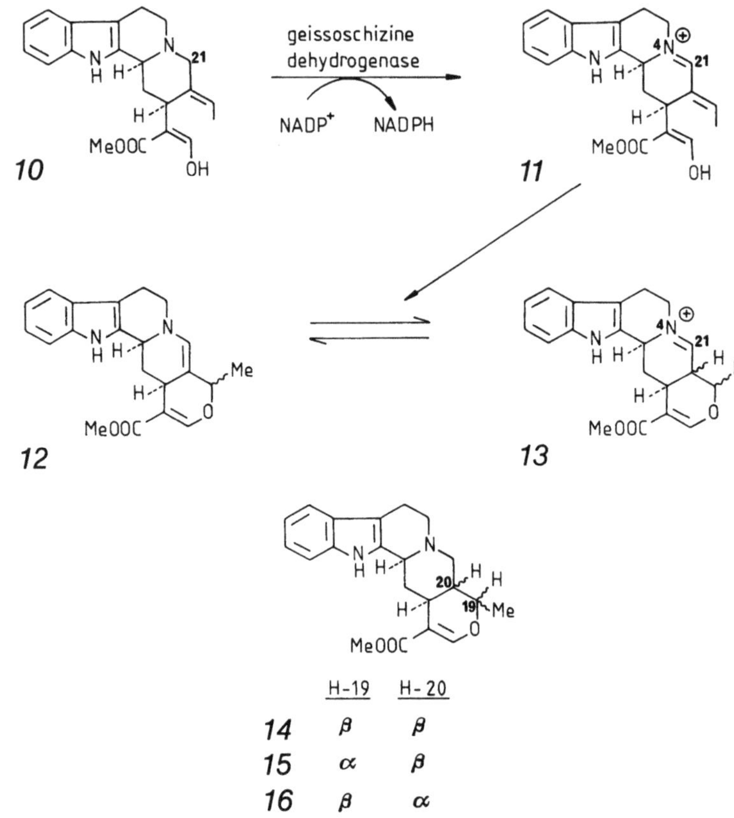

Fig. 4. The role of geissoschizine dehydrogenase in the formation of heteroyohimbine alkaloids in *Catharanthus roseus* cell suspension cultures

alkaloidal extracts by Extrelut columns and the identification and quantitation of alkaloids by means of HPLC has recently been described [36]. For the specific identification of a particular compound in plant cell cultures, the method which has proved to be most sensitive is that of radioimmunoassay. The detection of vindoline (*4*) by this technique has been described and it is sensitive within a range of 0.5–100 pmol (0.2–45 ng) [37].

In order to understand the formation and storage of alkaloids in cell suspensions of *C. roseus*, the localization of the alkaloids has been investigated. The alkaloids are stored within the vacuoles of particular cells which possess a vacuolar pH of 3 in contrast to "normal" cells which have a vacuolar pH of 5 [38]. The uptake and retention of tryptamine and alkaloids by cells in suspension have been the subject of several studies [39–41].

The manipulation of plant cell cultures to produce indole alkaloids continues to be an active area of research as exemplified by various studies on *C. roseus* cultures. Factors such as temperature [42] and light [43] have been investigated for their effect on alkaloid production. High sucrose levels stimulated the production of ajmalicine [44] whereas another study revealed that alkaloid synthesis continued with low phosphate levels but not with low sucrose levels [45]. Changes in media composition have been related to the activities of tryptophan decarboxylase and strictosidine synthase [46]. In an attempt to induce the production of dimeric alkaloids, the effect of several carotenoid-inducers have been investigated. Although ajmalicine and catharanthine production was increased by up to about 20% by concentrations of specific inducers, at levels of 5 ppm, there was no detection of dimers [47]. The omission of 2,4-D from media has resulted in notable increases of ajmalicine and serpentine [48].

Although *C. roseus* currently holds the centre of the stage for biosynthetic studies of indole alkaloids utilising plant cell cultures, there are a limited number of current studies of other genera from the Apocynaceae. Some twelve indole alkaloids have been isolated from cell suspension cultures of *Rauwolfia serpentina* [49]. The major alkaloid proved to be vomilenine (*23*) (ajmaline-type) which was present at much higher concentrations than in differentiated plants. Crude enzyme extracts prepared from the cell suspension cultures converted strictosidine to a series of yohimbine-, heteroyohimbine-, sarpagine- and ajmaline-type alkaloids. Polyneuridine aldehyde esterase (PNA-esterase), isolated and purified from cell suspensions of *R. serpentina*, specifically converts polyneuridine aldehyde (*17*) into 16-epivel-

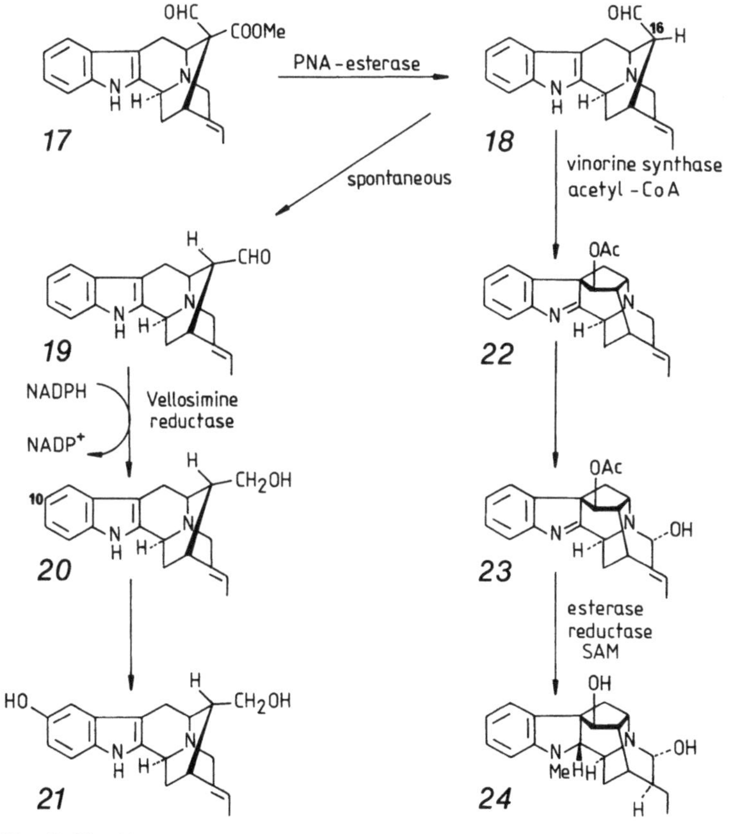

Fig. 5. The biosynthesis of sarpagine and ajmaline by means of enzymes isolated from *Rauwolfia serpentina* cell cultures

losimine (*18*) which is at the branch of the biogenetic routes leading to sarpagine- and ajmaline-type alkaloids [50] (Fig. 5). Enzyme extracts of *R. serpentina* cells converted polyneuridine aldehyde (*17*) to vellosimine (*19*) [51]. The same cultured cells were used as a source for the isolation of the next enzyme in the pathway, vellosimine reductase which specifically converts vellosimine (*19*) into 10-deoxy-sarpagine (*20*), the direct precursor of sarpagine (*21*) [52,53]. 16-Epi-vellosimine (*18*) would be anticipated to cyclise to deacetylvinorine and a further enzyme isolated from cell suspensions of *R. serpentina*, named vinorine synthase, is involved in the conversion of deacetylvinorine to vinorine (*22*) [54]. Ajmaline (*24*) is produced via volilenine (*23*) and the specific enzyme involved has been obtained from cell cultures of *R. serpentina* [55] (Fig. 5). Cell suspension cultures of *Tabernaemontana divaricata* and *T. iboga* have yielded a series of monoterpenoid indole alkaloids. The availability of strictosidine by means of immobilized strictosidine synthase [12] means that the role of strictosidine can be investigated by cell-free systems of these *Tabernaemontana* species. In particular, *T. iboga* should prove to be of interest since it produces only two major alkaloids, tubotaiwine (Strychnos-type) and conoflorine (quebrachamine-type) [56]. Further information on the biosynthesis of *Aspidosperma* alkaloids can be expected from studies of plant cell cultures because such alkaloids predominate in cultures of *Voacanga africana* [57].

2.2 Isoquinolines

Feeding experiments with callus cultures of *Stizolobium hasjoo* (Leguminosae) using DL-[β-^{14}C]-3,4-dihydroxyphenylalanine (DOPA) (*25*) [2-^{14}C] pyruvate and [2-^{14}C] acetate have resulted in the formation of labelled 3-carboxy-6,7-dihydroxy-1,2,3,4-tetrahydroisoquinoline (*27*) and 1-methyl-3-carboxy-6,7-dihydroxy-1,2,3,4-tetrahydroisoquinoline (*28*) [58,59]. Labelled DOPA was incorporated into both of these simple tetrahydroisoquinoline alkaloids whereas labelled acetate was incorporated only into (*27*) and labelled pyruvate was not incorporated into either alkaloid. It has been proposed that the formation of (*27*) and (*28*) proceeds via the α-carboxylic acid (*26*) (Fig. 6). The inital step in 1-benzyltetrahydroisoquinoline alkaloid biosynthesis has been assumed to be via a condensation of two aromatic units, both derived from tyrosine (*29*) namely dopamine (*30*) and 3,4-dihydroxyphenylacetaldehyde (*31*) to yield the first formed 1-benzyltetrahydroisoquinoline alkaloid, norlaudanosoline (*32*). Support for this pathway has been obtained by the use of cell cultures of various species of the Papaveraceae and of *Berberis* and *Thalictrum* [60,61]. An alternative biosynthetic pathway proposed for the formation of norlaudanosoline (*32*) is via

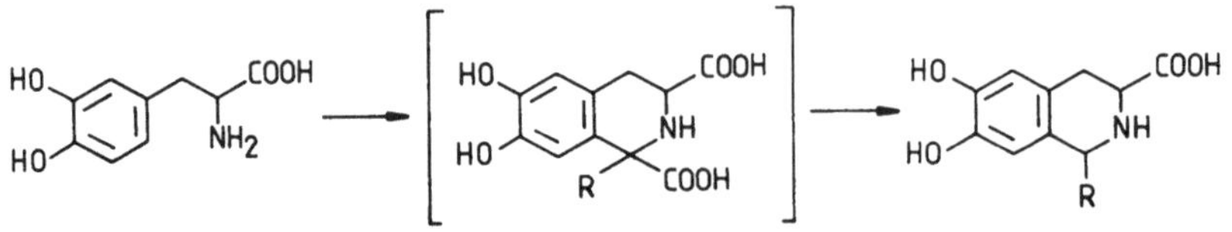

25 26 27 R=H
28 R=Me

Fig. 6. Formation of simple tetrahydroisoquinoline alkaloids by *Stizolobium* callus cultures

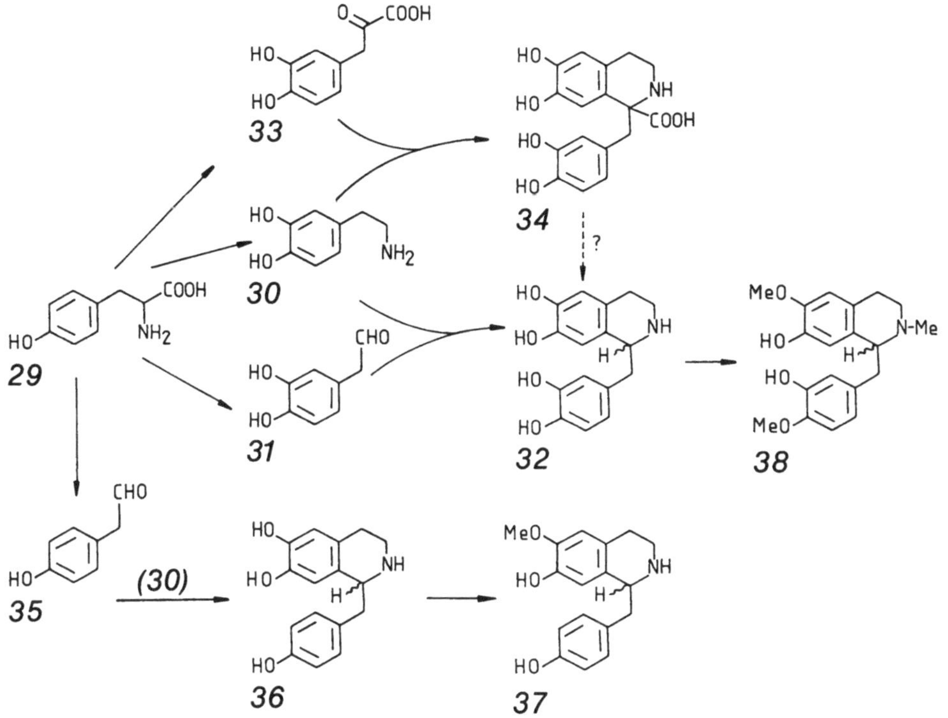

Fig. 7. Putative biosynthetic routes to the 1-benzyltetrahydroisoquinoline alkaloids

the condensation of dopamine (*30*) with 3,4-dihydroxyphenylpyruvate (*33*) to yield norlaudanosoline-1-carboxylic acid (*34*) which would then decarboxylate to form norlaudanosoline (*32*) (Fig. 7). Experiments with cell-free systems from suspension cultures have failed to produce evidence which supports the existence of this pathway [60,61].

The enzyme (S)-norlaudanosoline synthase has been isolated and characterized from cell cultures and its presence has been demonstrated in those species which produce 1-benzyltetrahydroisoquinoline alkaloids while it appears to be absent from other species which do not contain these alkaloids [60]. Four isoenzymes of (S)-norlaudanosoline synthase have been separated from cell cultures of *Eschscholtzia tenuifolia* and none of these enzymes are capable of catalyzing the condensation of dopamine (*30*) with 3,4-dihydroxyphenylpyruvate (*33*) clearly indicating that norlaudanosoline-1-carboxylic acid (*34*) is not directly involved in the pathway to 1-benzyltetrahydroisoquinoline alkaloids [61] (Fig. 7).

The K_M values for 3,4-dihydroxyphenylacetaldehyde (*31*) and for 4-hydroxyphenylacetaldehyde (*35*) proved to be almost identical indicating that the synthase is also responsible for the formation of demethylcoclaurine (*36*). This alkaloid is a precursor for coclaurine (*37*) which acts as a precursor for proaporphines of the crotonosine-type and for the aporphines, roemerine and mecambroline. It is thought that the predominant pathway (Fig. 7) in the biosynthesis of 1-benzyl-tetrahydroisoquinoline alkaloids leads to the formation of the (S)-enantiomers of norlaudanosoline (*32*) and of demethylcoclaurine (*36*) which act as intermediates of the morphinans with the (R)-configuration [61]. Methylation of norlaudanosoline (*32*) leads to the key intermediate reticuline (*38*) which is involved in the formation of various alkaloids including the morphinans, protoberberines, proaporphines and cularines.

Two new O-methyltransferases have been isolated from cell suspension cultures of *Argemone platyceras* [62]. S-adenosylmethionine: (R), (S) norlaudanosoline-6-0-methyltransferase catalyses the formation of 6-0-methylnorlaudanosoline (*39*) and to a minor extent 7-0-methylnorlaudanosoline (*40*) from (S)- and (R)-norlaudanosoline (*32*). This enzyme has been shown to be widely distributed in plants which contain 1-benzyltetrahydroisoquinoline alkaloids and is present in plant cell cultures and differentiated plants such as *Papaver somniferum*. S-adenosylmethionine: (6-0-methylnorlaudanosoline)-5′-0-methyltransferase catalyses the formation of nor-orientaline (*41*) from 6-0-methylnorlaudanosoline (*39*) and appears to be a highly specific enzyme [63] (Fig. 8).

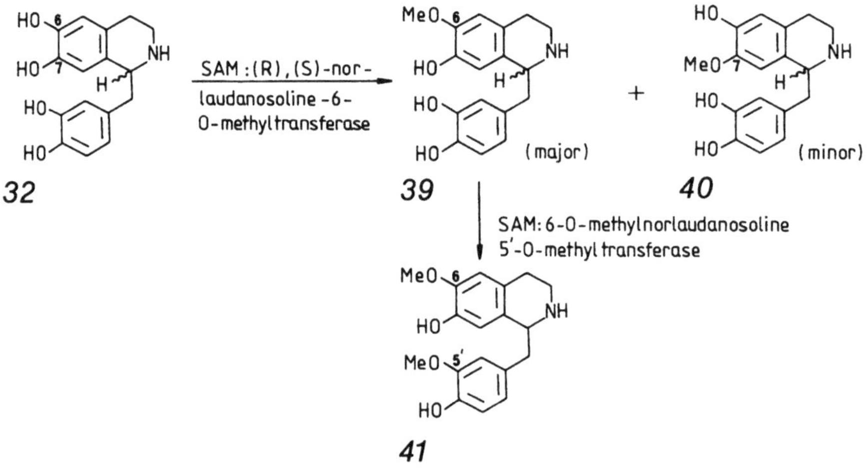

Fig. 8. Methyltransferases isolated from cell suspension cultures of *Argemone platyceras*

The formation of the protoberberine alkaloid skeleton and the pattern of oxygenation and methylation are intriguing biogenetic sequences. Incubation of (±)-reticuline (*38*) with [$^{14}CH_3$]-S-adenosylmethionine and a crude enzyme preparation from *Berberis aggregata* callus culture has resulted in the formation of labelled palmatine (*42*) and columbamine (*43*). However, similar experiments with the non-phenolic (±)-tetrahydropapaverine (*44*) failed to result in protoberberine production but the corresponding N-methylated alkaloid, laudanosine (*45*) was formed [64] (Fig. 9). The protoberberine alkaloid jatrorrhizine (*48*) is produced in high yields by some cell cultures of *Berberis* species. Doubly labelled 6-0-methyllaudanosoline (*46*) fed to *Berberis stolonifera* callus resulted in the formation of labelled berberine (*47*) and jatrorrhizine (*48*) [65] (Fig. 10). This finding suggests that jatrorrhizine is formed from berberine by reopening of the methylenedioxy group. Since [N-$^{14}CH_3$]-(±)-reticuline (*49*) was shown to be the favoured precursor for jatrorrhizine (*50*) in contrast to other phenolic 1-benzyltetrahydroisoquinoline alkaloids, including [N-$^{14}CH_3$]-protosinomenine (*51*), the results demonstrate that the inversion of A

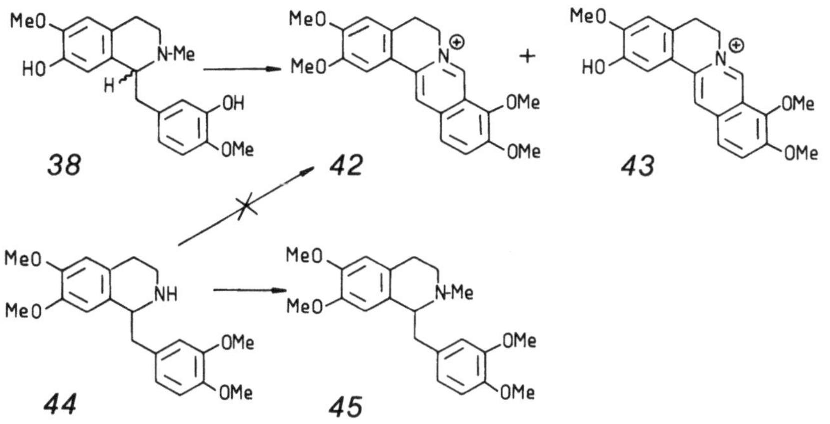

Fig. 9. Formation of protoberberine alkaloids from (±)-reticuline by callus cultures of *Berberis aggregata*

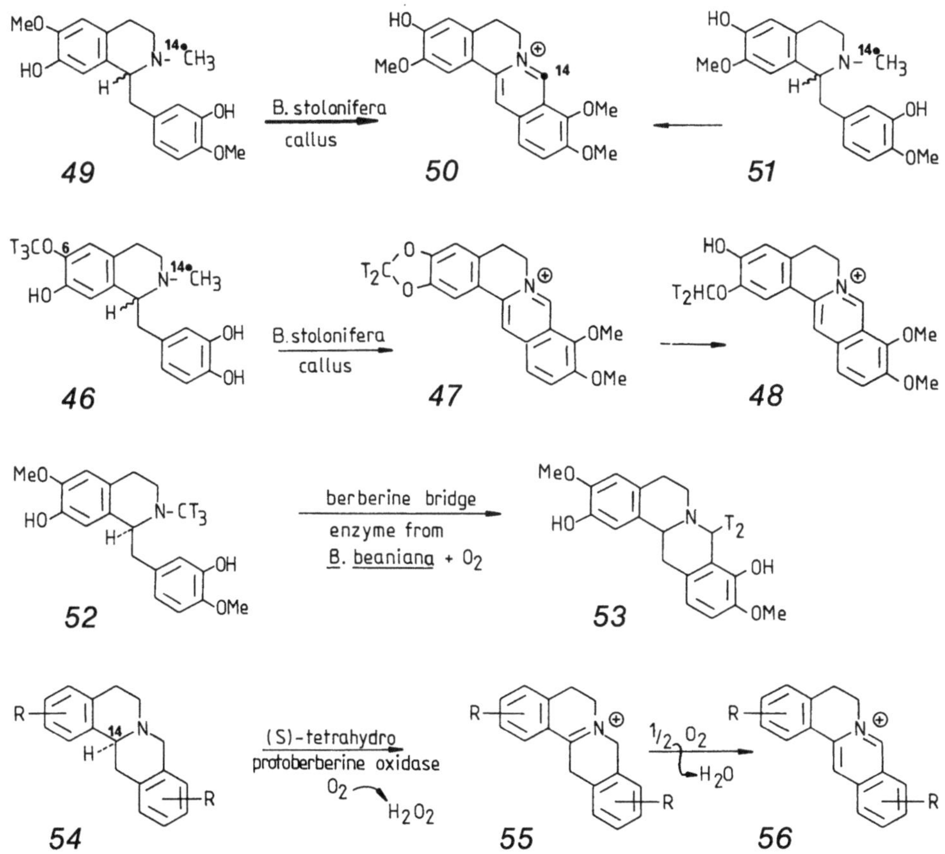

Fig. 10. Formation of protoberberine alkaloids in *Berberis* cell cultures and the "inversion" of methoxyl/hydroxyl substitution in ring A

ring methoxyl/hydroxyl substitution pattern, i.e. reticuline (*46*) → jatrorrhizine (*48*) proceeds via a methylene dioxy intermediate (*47*) (Fig. 10).

The berberine bridge enzyme has been isolated and purified from cell suspension cultures of *Berberis beaniana*. This enzyme is specific for substrates with the (S)-configuration and it catalyzes, in the presence of oxygen, the conversion of (S)-reticuline, (S)-protosinomenine and (S)-laudanosoline to the corresponding (S)-tetrahydroprotoberberines e.g. (S)-reticuline (*52*) → (S)-scoulerine (*53*) [66] (Fig. 10). Although this enzyme has been investigated previously in a cell-free system of *Macleaya microcarpa* cell cultures, the mechanism of conversions with regard to the proposed intermediacy of reticuline-N-oxide, the participation of oxygen and the steric course of the reaction remained unclear. The experiments with the enzyme isolated from *B. beaniana* indicate that reticuline *N*-oxide does not serve as a substrate, that oxygen is involved and that only the (S) configuration precursors act as substrates [66].

A new flavin enzyme has been isolated from cell cultures of several *Berberis* species. In the presence of oxygen, this enzyme catalyses the oxidation of (S)-tetrahydroprotoberberines (*54*) to protoberberines (*56*) via the 7,14-dehydroberberine intermediate (*55*). This (S)-tetrahydroprotoberberine oxidase is the final enzyme involved in the biosynthesis of protoberberine alkaloids [67] (Fig. 10).

Suspension cultures of *Papaver somniferum* have been shown to produce the labelled morphinan alkaloids, thebaine, codeine and morphine when fed with ^{3}H-tyrosine [68].

Protein synthesis inhibitors, cycloheximide, puromycin (PUM) and actinomycin (ATM) in low concentrations promoted alkaloid synthesis and in particular low concentrations of PUM and ATM increased the accumulation of codeine but not of morphine. Cell suspension cultures of *P. somniferum* cv Marianne are capable of transforming thebaine (*57*) to neopine (*58*) whereas codeine, neopine, papaverine and laudanosoline were not metabolized by these cultures [69] (Fig. 11).

57 58

Fig. 11. Conversion of thebaine to neopine by cell suspension cultures of *Papaver somniferum*

2.3 Quinolizidines

The administration of non-quinolizidine alkaloids such as papaverine and coniine to cell suspension cultures of *Lupinus polyphyllus* has resulted in an increase in quinolizidine alkaloid production [70]. Surprisingly, similar methods of induction have produced quinolizidine alkaloids in species which either produce other alkaloids or no alkaloids, viz. *Conium maculatum*, *Daucus carota*, *Atropa belladonna*, *Chenopodium rubrum*, *Spinacia oleracea* and *Symphytum officinale*. These results suggest that genes which are responsible for quinolizidine alkaloid production are not restricted to the Leguminosae but are widely distributed in higher plants.

Cell suspension cultures of *Lupinus polyphyllus* have been used to study the biochemistry of alkaloid biosynthesis. By means of labelled precursors or intermediates introduced into suspensions, which have low alkaloid content, it has proved possible to determine biosynthetic pathways. Although these sequences may deviate from those occurring in intact plants, the use of cell suspensions does enable an understanding of the likely biosynthetic relationships. Cadaverine (*59*) has been converted within the cell suspensions to 17-oxosparteine (*60*), 12,13-dehydro-17-oxosparteine (*61*), lupanine (*62*), sparteine (*63*), and 12,13-dehydrosparteine (*64*) [71]. The possible sequence of quinolizidine alkaloid transformation in *Lupinus* cell suspension cultures is summarized in Fig. 12. Quinolizidine alkaloid production in *L. polyphyllus* is localized in leaf chloroplasts and it is not surprising that alkaloid production is higher in photomixotrophic cell suspension when compared with chlorophyll-free heterotrophic cultures [72]. Lupanine is a key alkaloid for a series of quinolizidines and the major alkaloid produced by cell suspension cultures [73] although its concentration is some 200 times lower than in differentiated lupins. The use of GLC-MS techniques has facilitated the identification of quinolizidine alkaloids in cell suspensions of *L. polyphyllus* and although lupanine represents 97% of the total alkaloid, it has proved possible to identify sparteine, 13-angeloyloxylupanine and 13-tigoyloxylupanine. Lupanine, 13-*cis*- and 13-*trans*-cinnamoyloxylupanine were detected in the spent cell culture medium. The corresponding differentiated plants yielded some 18 quinolizidine alkaloids [74].

The main alkaloid produced by cell suspensions of *Baptisia australis* is lupanine [75]. When [^{14}C]-lupanine was fed to *B. leucopheya* plants a series of radioactive alkaloids were formed. As a direct result of this study, and of previous work, it was possible to propose a biosynthetic sequence from lysine to cadaverine (*59*) and then

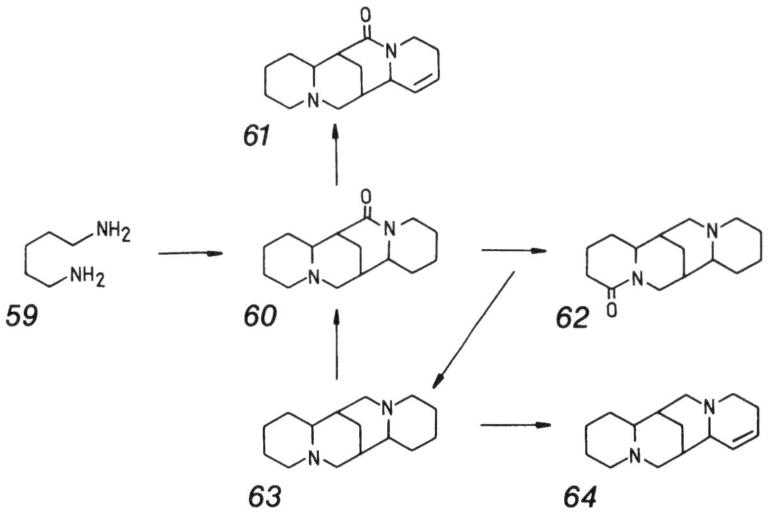

Fig. 12. Possible sequence of quinolizidine alkaloid transformation in cell suspension cultures of *Lupinus polyphyllus*

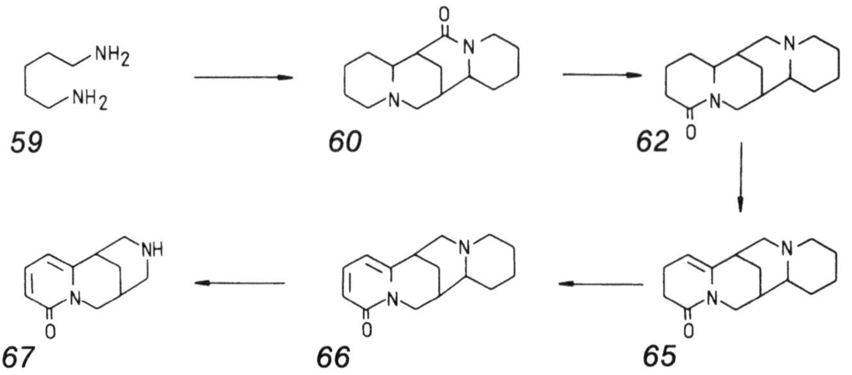

Fig. 13. Proposed biogenetic pathway of quinolizine alkaloids in *Baptisia* species

to the alkaloids, 17-oxosparteine (*60*), lupanine (*62*), 5,6-dehydrolupanine (*65*), anagyrine (*66*) and cytisine (*67*) (Fig. 13).

The patterns of quinolizidine alkaloids in cell cultures of ten species of Leguminosae have been analyzed by GLC-MS and compared with the alkaloids present in the leaves of the respective plants [76]. Lupanine proved to be the major alkaloid in all ten cell suspension cultures. Although α-pyridones are the major alkaloids in the genera *Cytisus*, *Genista*, *Laburnum* and *Sophora* they did not accumulate in the cell cultures. These findings support the view that the pathway leading to lupanine is the basic pathway of quinolizidine alkaloid biosynthesis and that the other alkaloids are derived from lupanine.

17-Oxosparteine synthase, the enzyme responsible for the conversion of cadaverine (*59*) to 17-oxosparteine (*60*) has been demonstrated to be present in chloroplasts and in cell suspensions of *Lupinus polyphyllus* [77,78]. Lysine decarboxylase, which converts lysine to cadaverine (*59*), is widespread in higher plants and in *L. polyphyllus*, it is also associated with the chloroplasts. This enzyme has been detected in suspension cultures of five different genera which elaborate quinolizidine alkaloids [79].

2.4 Tropanes

Newly induced callus cultures of *Atropa belladonna*, *Datura stramonium* and *Hyoscyamus niger* have been examined for their ability to produce tropanes and only the latter plant yielded alkaloids in this study. Hyoscyamine (*68*) and scopolamine (*70*) were identified by means of TLC, GLC and GLC-MS [80]. However, other workers have produced tropane alkaloids from cultures of *A. belladonna* roots and atropine has been reported as the major alkaloid. Alkaloid production was stimulated by supplementation of the medium by arginine, ornithine, proline and tyrosine [81]. The effects of pH and of media composition have been investigated for the production of tropane alkaloid by suspension cultures of *Hyoscyamus muticus*. At a pH of 3.5 the maximum yield of alkaloid per dry weight of cells rose to 0.15%. Apparently the production of alkaloids was independent of morphogenesis and this finding contrasts with that of other workers who have reported that alkaloid production in *A. belladonna* and in *Scopolia parviflora* is coupled with root organization [82].

Callus from a hybrid of *Duboisia leichardtii* and *D. myoporoides* has been supplemented with hyoscyamine (*68*) and the metabolites identified as 6-hydroxyhyoscyamine (*69*) and scopolamine (*70*). These results substantiate the proposal that scopolamine biosynthesis in *Duboisia* species appears to be via the hyoscyamine 6-hydroxyhyoscyamine pathway [83]. (Fig. 14). Tissue cultures of *Datura innoxia* possess the ability to esterify tropine and tropic acid whereas similar cultures prepared from *A. belladonna*, *H. niger*, and *S. lurida* were unable to facilitate this esterfication [84].

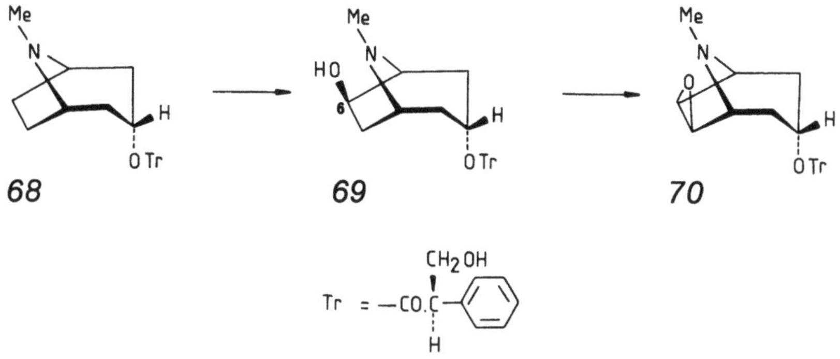

Fig. 14. Conversion of hyoscyamine (68) to 6-hydroxyhyoscyamine (69) and scopolamine (70) by callus cultures of a *Duboisia* hybrid

Cell cultures of *D. innoxia* can acetylate scopine, scopoline, pseudotropine and tropine [85].

Root cultures of *D. stramonium* have yielded an NADPH-specific tropine dehydrogenase which catalyzes the oxidation of tropane-3-α-ols but not tropane-3-β-ols to the corresponding ketone [86]. These workers make the assumption that this dehydrogenase is in some way involved in the biosynthesis of tropane alkaloids, even though the accepted biosynthetic pathway is from tropinone to tropane-3-α-ol and not vice versa. It has not proved possible to isolate this enzyme from root cultures of *A. belladonna* plants or of *D. stramonium* [86].

2.5 Nicotine

The biosynthesis of tobacco alkaloids has recently been the subject of an extensive review article [87] and it would appear that little use has been made of cell cultures for biosynthetic studies of nicotine and related alkaloids. Callus cultures of four burley tobacco genotypes were supplemented with a number of possible intermediates in the biosynthesis of nicotine e.g. pyridine, pyrrolidine, nicotinic acid, putrescine, ornithine. Differences in nicotine concentrations among the callus cultures of the four alkaloid genotypes were not attributed to biosynthetic block in the formation of the pyridine or pyrrolidine ring of nicotine [88]. In another study involving *Nicotiana tabacum* callus, the activities of phenylalanine aminotransferase, ornithine aminotransferase, ornithine transcarbamylase and arginase, have been correlated with nicotine biogenesis [89]. Some callus tissues of *N. tabacum* fail to produce any nicotine and it has been proposed that this may be due to lack of putrescine N-methyltransferase [90]. The formation of putrescine has been investigated in cell suspensions of *N. tabacum* which contained either high or low levels of cinnamolyputrescines. Labelled arginine and ornithine were equally well incorporated into caffeoyl and feruloylputresine [91]. The level of cinnamoylputrescines in cell suspensions of *N. tabacum* has been correlated with phosphate concentration in the media and with the activity levels of specific enzymes [92-97].

2.6 Acridones

Anthranilic acid (*71*) is specifically incorporated into rutacridone (*73*) by callus and cell suspension cultures of *Ruta graveolens* [98]. Cell-free extracts from cell suspensions of *R. graveolens* have been shown to catalyze the N-methylation of anthranilic acid (*71*) to form N-methylanthranilic acid (*72*) using S-adenosyl-L-methionine as methyl donor. This N-methylation is the first specific step in the biosynthesis of acridone alkaloids such as rutacridone (*73*) [99]. [1-^{13}C]-[2-^{13}C]- and [1,2-$^{13}C_2$]-acetate fed to cell suspensions of a cell line of *R. graveolens* have resulted in ^{13}C labelling of ring C in rutacridone [100] (Fig. 15).

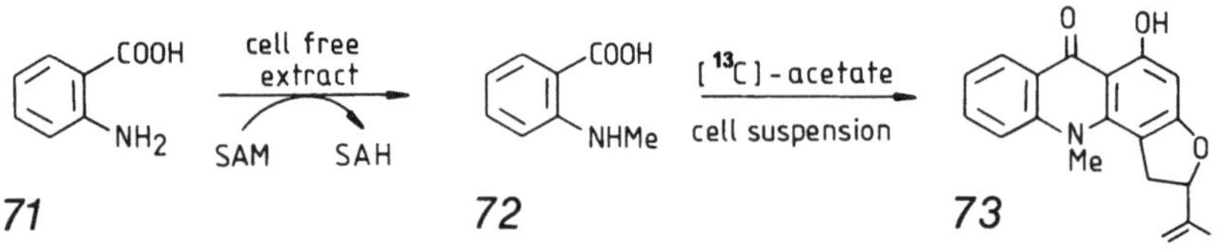

Fig. 15. First pathway-specific reaction of rutacridone (73) biosynthesis in *Ruta graveolens*

2.7 Steroids

[^{3}H]-solanidine is rapidly glycosylated by cell suspensions of *Solanum tuberosum* to produce 3-β-O-glycosylsolanidine (γ-chaconine) and 3-β-O-glycosyl (glucosyl) solanidine [101]. The biosynthesis of steroidal alkaloids in the Solanaceae has been extensively reviewed and it has been pointed out that experiments with plant cell culture techniques have provided useful data [102].

2.8 Purines

The production of theobromine and caffeine in cell suspensions of *Coffea arabica* has been studied and it has been shown that the rate of transformation of theobromine to caffeine is highest during the log phase of growth [103]. Tracer experiments with [2-^{14}C] caffeine demonstrated that there was only slow degradation of this alkaloid [103]. Cell-free extracts of coffee fruits and of callus have been prepared and the activities of S-adenosyl-L-methionine N-methyltransferase and

N-methyl-N-nucleoside hydrolase have been compared [104]. It has been proposed that callus cultures of *C. arabica* form caffeine via the same biosynthetic route which takes place in coffee fruits and in tea leaves [105]. The overall metabolism of [8-^{14}C] adenine, [8-^{14}C] adenosine, [8-^{14}C] guanine, [8-^{14}C] hypoxanthine and [8-^{14}C] inosine in cell suspensions of *Catharanthus roseus* has been studied [106].

3 Phenols

3.1 Simple Phenols

The sequence of reactions converting phenylalanine into the CoA ester derivatives of substituted cinnamic acids ("phenylpropanoid metabolism") has been reviewed in the context of flavonoid biosynthesis [107,108]. A series of enzymes, which catalyze these reactions, have been isolated and characterized, including phenyl-ammonia lyase (PAL), cinnamate 4-hydroxylase and 4-coumarate: CoA ligase. *Trans* cinnamic acid, the product resulting from PAL activity has a regulatory effect on the enzyme and so have the hydroxylated cinnamates, p-coumaric and caffeic acids [109]. Cell suspension cultures of apple (*Pyrus malus*) accumulate hydroxy-cinnamoyl esters including chlorogenic acid, feruloylquinic acid, sinapoyl glucose, feruloyl glucose and p-coumaroyl glucose. It has been demonstrated that as these cultures age, there is a significant increase in the accumulation of feruloyl and sinapoyl derivatives and a corresponding decrease in the chlorogenic acid content. Intensive phenolic synthesis appeared to occur during active cell division of the cultures and the activity of the enzymes involved in phenylpropanoid biosynthesis correlated with the sequence of phenolic synthesis. The O-methylation of chloro-genic acid suggests a possible route for the biosynthesis of feruloylquinic acid while the patterns of glucosyl transferase activity indicates that glucosylation is a terminal step in the biosynthesis of phenolic compounds [110,111].

Hydroxyphenylethanol glycosides accumulate at up to 16% of the dry weight of cells in suspension cultures of *Syringia vulgaris*. Tyrosine and tyramine are efficient biosynthetic precursors of 4-hydroxy- and 3,4-dihydroxyphenylethanol moieties of these glycosides [112]. Cell suspensions of *Lithospermum erythrorhizon* and *Gardenia jasminoides* glucosylate salicyl alcohol to form the phenolic glycoside salicin whereas cultures of *Datura innoxia*, *Nicotiana tabacum*, *Duboisia myoporoides*. *Catharanthus roseus* and *Bupleurum falcatum* form the corresponding alcoholic glycoside, isosalicin, when they are fed with salicyl alcohol [113].

3.2 Flavonoids

It is well established that the first formed intermediate in flavonoid biosynthesis is a chalcone (*74*) on which ring A is derived by head to tail condensation of three acetate units and the remaining C_6—C_3 unit is derived via phenylalanine. The chalcone (*74*) isomerises to flavanone (*75*) which acts as an intermediate in the production of flavone (*76*), flavonol (*77*), anthocyanin (*78*), isoflavone (*79*),

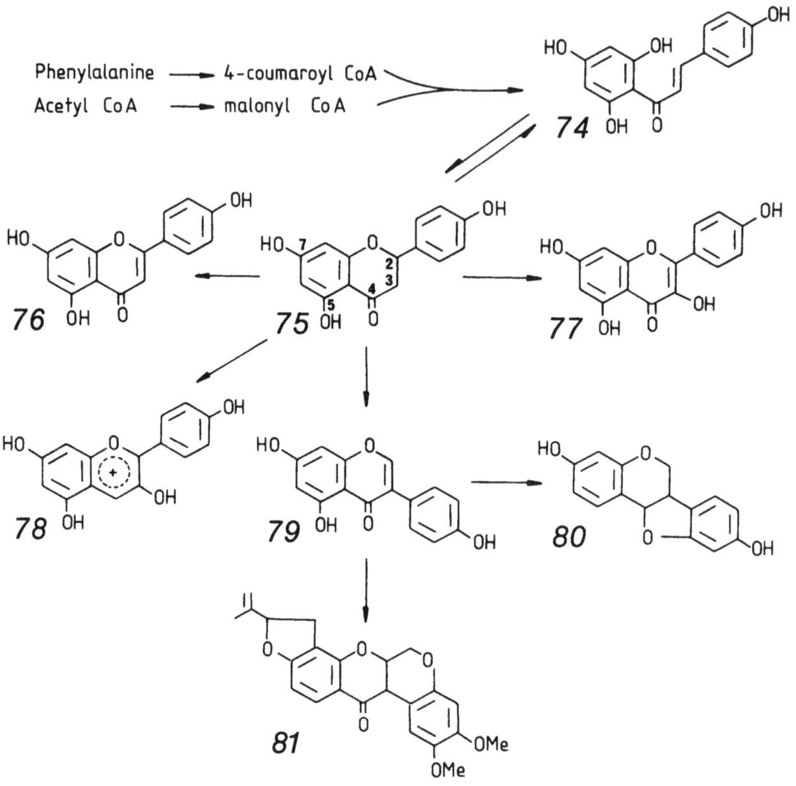

Fig. 16. Chalcone (74) — flavanone (75) as the common intermediates in the biosynthesis of flavonoids

pterocarpan (*80*) and rotenone (*81*) molecules (Fig. 16)[107,108]. A strong correlation exists between flavonoid production and the expression of PAL activity in developing plant tissues or after the treatment of plant tissues or plant cell cultures with light, microorganisms, microbial products, hormones or other chemicals[107,114,115]. Phytochrome and a blue-light receptor are involved in UV-B (280–320 nm) induced flavonoid biosynthesis in parsley (*Petroselinum hortense*) cell suspension cultures[116]. The dependency of the UV-effect on the amont of active far red absorbing form of cytochrome (PFr) has been investigated. The effects of continuous irradiations with red, far red and blue light have been compared with the effects of multiple pulses of the same wavelengths to find out whether a classical high irradiance response (HIR) was involved. The results of these experiments suggest that flavonoid synthesis in cell cultures of parsley is controlled by three photoreceptors. Excitation of the UV-B photoreceptor is obligatory but expression of this effect is dependent on PFr and furthermore, flavonoid synthesis is also stimulated by a separate blue light photoreceptor. It has been shown that the de novo synthesis of PAL is phytochrome, and therefore light, mediated[117].

In addition to the formation of the 4-hydroxylated flavonoids from cinnamoyl CoA, further hydroxylations and methylations commonly occur at positions 3 and 5 but to date it is the methyl-transferases which have received more study than the hydrolases [118]. The role of the phenolases which may catalyse hydroxylations of phenylpropanoids and of flavonoids has not been fully established [107]. The acetyl-CoA carboxylase which is involved in flavonoid biosynthesis in parsley has been purified and characterized to some extent [119].

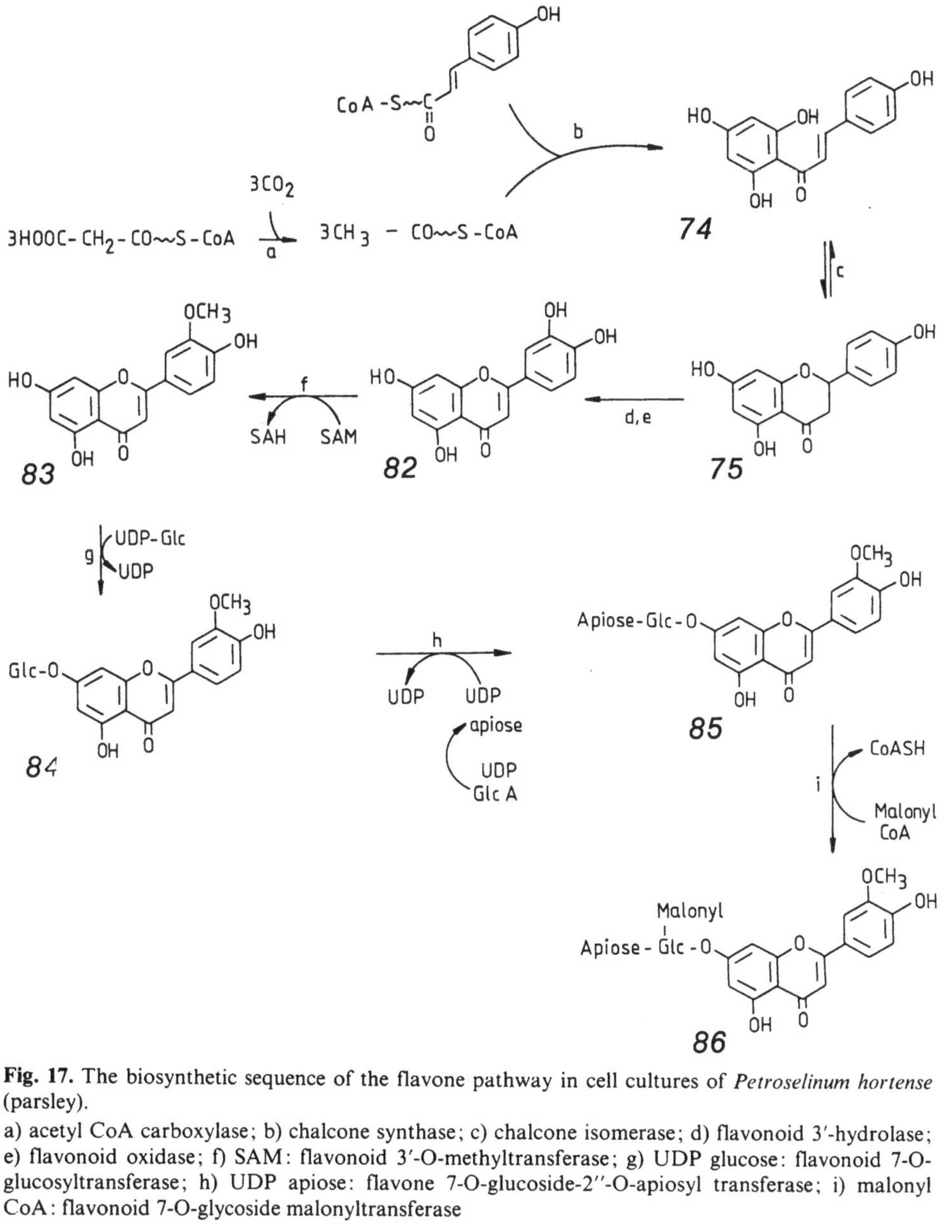

Fig. 17. The biosynthetic sequence of the flavone pathway in cell cultures of *Petroselinum hortense* (parsley).

a) acetyl CoA carboxylase; b) chalcone synthase; c) chalcone isomerase; d) flavonoid 3′-hydrolase; e) flavonoid oxidase; f) SAM: flavonoid 3′-O-methyltransferase; g) UDP glucose: flavonoid 7-O-glucosyltransferase; h) UDP apiose: flavone 7-O-glucoside-2″-O-apiosyl transferase; i) malonyl CoA: flavonoid 7-O-glycoside malonyltransferase

Incubation of parsley cell suspensions with [1,2,$^{13}C_2$] acetate followed by ^{13}C FT NMR studies of isolated apigenin (flavone) and kaempferol (flavonol) has confirmed that ^{13}C enrichment, due to acetate incorporation, is primarily in ring A [120]. Coupling patterns between adjacent ^{13}C atoms of ring A indicate that the cyclization direction of ring A is random in both compounds. Randomization of apigenin labelling could have occurred chemically through the opening of the pyrone ring under the acidic conditions used for glycoside hydrolysis but randomization in labelling of the more stable flavonol must have occurred biosynthetically. These results support the view that a chalcone (*74*) is an intermediate in flavonoid biosynthesis.

The chalcone/flavonoid isomers are the central intermediates in the synthesis of all flavonoids (Figs. 16 and 17) [108,121]. Compelling evidence has now been presented that the chalcone (*74*) is the intermediate product of the synthase which catalyzes the condensation of acyl residues from one molecule of 4-coumaryl-CoA and three molecules of malonyl CoA [122,123]. In a second step, chalcone isomerase catalyzes the stereospecific formation of the (2S)- flavanone (*75*) from the corresponding chalcone (*74*) (Fig. 17) [124]. Regulation of chalcone synthase has been further studied by investigations of the UV-induction of chalcone synthase m-RNA in parsley cell suspension cultures. DNA's complementary to Poly(A)$^+$ in RNA's from UV-irradiated cell suspension cultures of parsley were inserted into the cloning vehicle pBR 322 and used to transform *Eschericha coli* strain RRl. A clone containing a DNA complementary to chalcone synthase mRNA was identified by hybrid-selected and hybrid-arrested translation. Large and rapid changes in the amount of chalcone synthase m-RNA in response to irradiation of the cells was detected by RNA blot hybridization experiments. The pattern of changes coincides with that previously determined for the rate of chalcone synthase synthesis as measured either in vivo or with polyribosomal m-RNA in vitro [125].

A few studies with cell-free extracts have demonstrated the conversion to flavanones (*75*) to flavones (*76*) and the C-3 hydroxylation of flavanones to dihydroflavonols [108]. The concentrations of flavone (*76*) and flavonol (*77*) glycosides in cell suspensions of parsley, which had been irradiated with light, were similar to the amounts found in the leaves and seeds of whole plants. Hence parsley cell suspension cultures have proved useful for studying the biosynthesis of flavones (Fig. 17) and flavonols. A soluble enzyme isolated from these cell suspensions catalyzes the conversion of flavanone to flavone, dihydroflavonol to flavonol [126]. As a result of these studies, it has been postulated that flavone (*76*) and flavonol (*77*) synthesis from flavanone (*75*) proceeds via 2-hydroxy- and 2,3-dihydroxy-flavanone with subsequent dehydration. The microsomal fraction of the parsley cells contains an NADPH-dependent flavanone-3′-hydroxylase.

It is most likely that glycosylation occurs subsequent to all other substitutions and modifications of the flavonoid ring system [107,108]. The branch chain sugar, apiose, is found only in the flavone glycosides and not in other glycosides of the parsley cell cultures. The enzyme responsible for the apiosyl transfer from UDP-apiose has been shown to require mRNA dependent regulation [127].

Malonylation of the sugar moieties is the last step in the biosynthesis of flavonoid glycosides in cultured parsley cells and several malonyl-flavone and -flavonol glycosides have been isolated [128,129]. A summary of the biosynthetic sequence determined from studies of parsley cell cultures from flavanone (*75*) intermediate via hydroxylation, methylation, glucosylation and malonylation in the flavone series is given with the enzymes which are involved in Fig. 17 (*75* → *82* → *86*).

Echinatin (*90*) is a biosynthetically unique retrochalcone in which the origins of the two aromatic rings are the reverse of that which occurs in normal flavonoids (Fig. 18). The dibenzoylmethane, licodione (*88*) has been demonstrated to be an obligate intermediate in the biosynthesis of echinatin (*90*) in cultured cells of *Glycyrrhiza echinata* [130,131]. This view has been supported by the detection of an 0-methyl transferase which catalyzes the specific 2′-0-methylation of licodione in the cultured cells. The incorporation of ^{3}H-isoliquiritigenin (*87*) into (*89*) has led to the suggestion that the biosynthetic course does involve the conversion of an unsaturated ketone unit. Isolation of the labelled licodione (*88*) suggests that it may be an intermediate in this process (Fig. 18) [132,133].

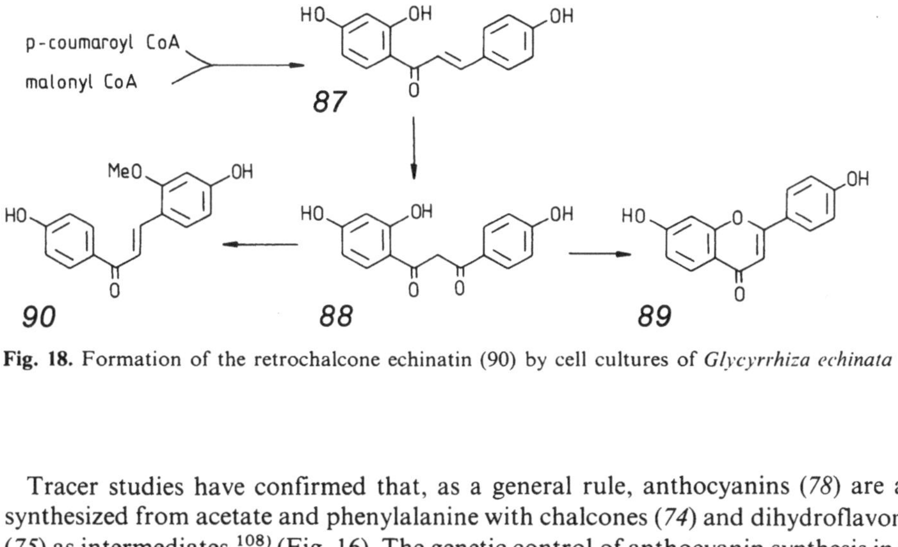

Fig. 18. Formation of the retrochalcone echinatin (90) by cell cultures of *Glycyrrhiza echinata*

Tracer studies have confirmed that, as a general rule, anthocyanins (*78*) are also synthesized from acetate and phenylalanine with chalcones (*74*) and dihydroflavonols (*75*) as intermediates [108] (Fig. 16). The genetic control of anthocyanin synthesis in cell cultures of *Matthiola incana* (Cruciferae) has been studied and it has been shown that callus of the cyanic line produced genotype specific anthocyanin patterns whereas callus cultures of the acyanic line produced no anthocyanins [134].

Proanthocyanidins ('condensed tannins') and their potential precursors have been isolated from cell suspension cultures of Douglas fir (*Pseudotsuga menziesii*) and the total proanthocyanidin content proved to be either equal to, or greater than, that found in mature needles [135]. The major monomer was identified as catechin and the major dimer as epicatechin-catechin. L-[U^{14}C]-phenylalanine fed to cell suspension cultures was incorporated into (+)-catechin (*91*) and (—)-epicatechin (*92*) and into a series of procyanidins of increasing molecular weight, e.g., procyanidin (*93*). Asymmetric labelling of dimers and polymers was demonstrated and it was concluded that the carbon-cation with the 2,3-*cis* stereochemistry of (—)-epicatechin (*92*) was formed more rapidly than was that of the 2,3-*trans* type of (+)-catechin (*91*)

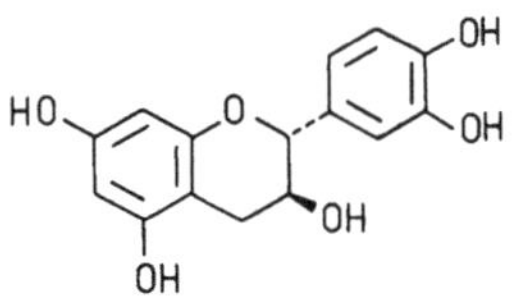

91

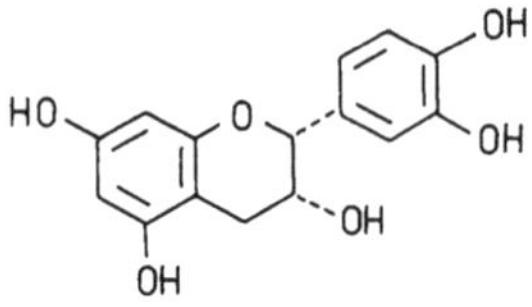

92

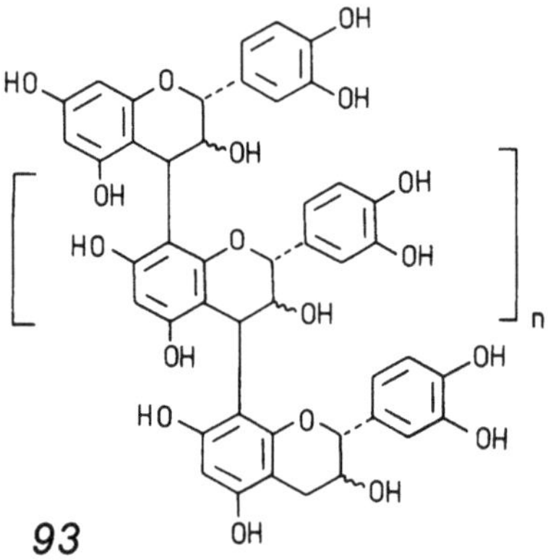

93

Fig. 19. Precursors of proanthocyanidins from cell suspension cultures of Douglas fir *(Pseudotsuga menziesii)*

(Fig. 19) [136]. Clear evidence exists which also demonstrated that isoflavonoids (*79*) are derived via chalcones (*74*) and flavanone (*75*) (Fig. 16) although the actual mechanism of the aryl migration of ring B from C-2 to C-3 is not fully understood. The accumulation of glyceollin in cell suspensions of soybean (*Glycine max*) coincides with large increases in the activities of PAL, cinnamate-4-hydrolase, 4-coumarate: CoA ligase, chalcone synthase and chalcone isomerase [108].

The formation of pterocarpans (*80*) has also been investigated by using plant cell cultures. The presence of dimethylallyl transferase has been demonstrated in cell free extracts obtained from soybean cell suspensions [137, 138]. This enzyme catalyzes the formation of 2-dimethylallytrihydroxypterocarpan (major product) and the corresponding 4-substituted pterocarpan. Rotenoids (*81*) which are derived via isoflavones (*79*) have been produced by cell cultures of *Derris elliptica* and although rotenoid biosynthesis increased during subculturing over a period of four months, it thereafter decreased and was finally lost. Rotenone (*81*) and deguelin were identified by GC—MS from callus containing imperfectly developed rootlets [139].

3.3 Naphthoquinones and Anthraquinones

As a result of earlier investigations with cell suspension cultures of *Galium mollugo*, the biosynthesis of quinones, naphthoquinones and anthraquinones has received further attention [140–143]. Studies based on the constituents of callus cultures of *Catalpa ovata* have indicated that 4-(2′-carboxyphenyl)-4-oxobutanoic acid (*94*) is the precursor of 2-carboxy-4-oxo-1-tetralone (COT) (*95*), an intermediate of several prenylated naphthoquinones. The main route of biosynthesis is via COT (*95*) to (2S)-prenyl COT (*96*) and (2R)-catalponone (*97*) while a subsidiary route proceeds via 2-carboxy-4-hydroxy-1-tetralone (CHT) (*98*) to prenyl CHT (*99*) and then to catalponol (*100*) (Fig. 20) [144, 145].

Callus cultures of *Echium lycopsis* have been shown to produce a mixture of red pigments which consists of five esterified derivatives of 5,8-dihydroxy-2-(1-hydroxy-4-methyl-3-pentenyl)-1-4-naphthoquinone. The cultures produced both the R-form (shikonin, *101*) and the (S)-form (alkannin, *102*) in various ratios depending upon the esterified derivative (Fig. 21). In contrast, *Lithospermum* cultures produced mainly the R-form prenylnaphthoquinones [146]. Shikonin (*101*), which is used pharmaceutically and as a dye, is now produced industrially by large scale plant cell culture in Japan [147].

Three unusually prenylated naphthoquinones (*103*, *104*) and (*105*) have been isolated from *Streptocarpus dunnii* and its cell cultures. The anthraquinones (*106*) and (*107*) have been obtained from the same rubiaceous plant (Fig. 21) [148].

Administration of ^{13}C- and ^{3}H-labelled precursors of *S. dunnii* cell cultures has demonstrated that naphthoquinones are formed through a unique prenylation mode [149]. Callus cultures of *Rumex alpinus* root produce hydroxy-anthraquinones, -dianthrones and naphthalenes which vary according to the hormones used in the culture medium. In the presence of 2,4-D the cultures produced the anthraquinones, chrysophanol (*108*), physcion (*109*) and emodin (*110*), dianthrones of chrysophanol and physcion and their heterodianthrones, the monoglucoside of chrysophanol

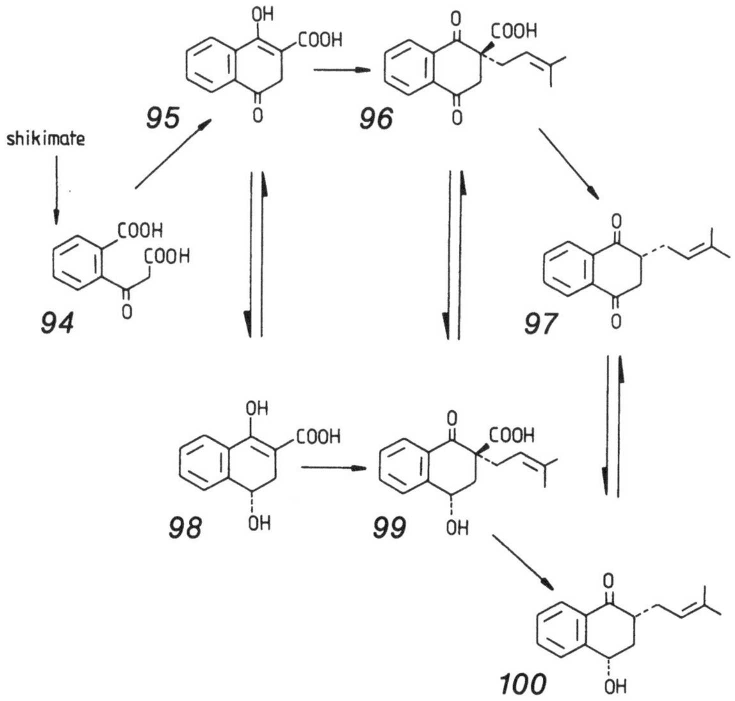

Fig. 20. Biosynthesis of prenylnaphthoquinones in callus cultures of *Catalpa ovata*

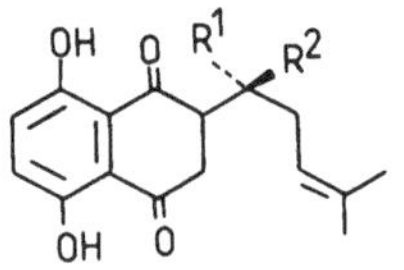

101 $R^1 = OH, R^2 = H$
102 $R^1 = H, R^2 = OH$

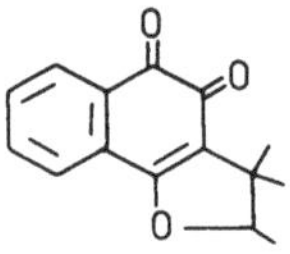

103

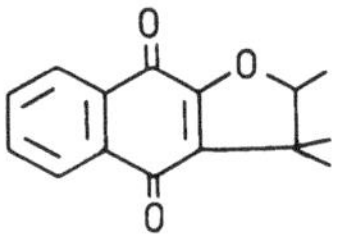

104

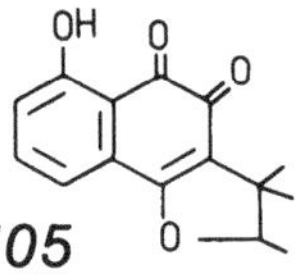

105

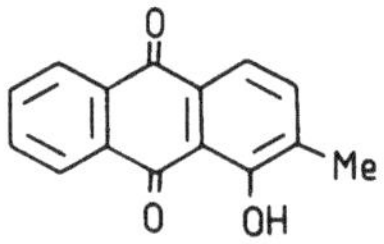

106

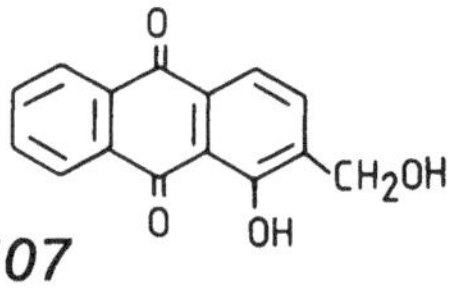

107

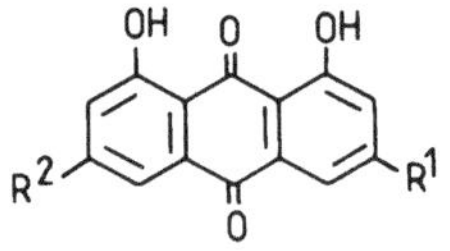

108 $R^1 = Me, R^2 = H$
109 $R^1 = Me, R^2 = OMe$
110 $R^1 = OH, R^2 = Me$

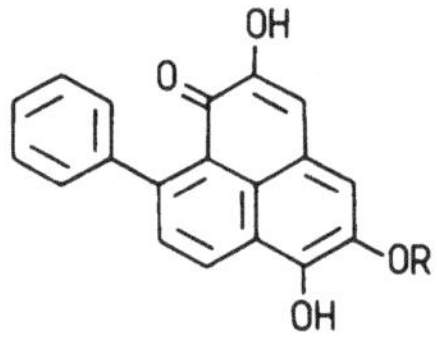

111 R = H
112 R = Me

Fig. 21. Examples of prenylnaphthoquinones and anthraquinones produced by plant cell cultures

together with the naphthalene-1-8-diols, nepodin, nepodin monoglucoside and methoxynepodin (Fig. 21) [150]. Anthraquinone formation in *Aloe saponaria* callus cultures is beneficially affected by light, although it has an inhibitory effect on the accumulation of the tetrahydroanthracene glucosides which are normally found in the subterranean organs of the differentiated plant. Light irradiation of these cultures accelerated the metabolism of tetrahydroanthracene glucosides to form anthraquinone glucosides such as those of chrysophanol (*108*) [151].

High levels of anthraquinones interfere with the measurement of some enzymes. A method has been developed for the determination of enzyme activities using cell suspensions of *Galium mollugo*. Under normal circumstances, the anthraquinones are located within the cell vacuole where they are spacially separated from the bulk of protein but homogenization of the cells results in the enzymes being inactivated by the anthraquinones [152]. ^{13}C-Labelled 2-succinylbenzoate has been incorporated into anthraquinones and naphthoquinones produced by cell suspension cultures of *G. mollugo* [143]. Feeding experiments with *Streptocarpus dunnii* cell cultures have provided evidence to indicate that anthraquinones are derived from naphthoquinones via prenylated intermediates [149].

The formation of 9-phenylphenalenone pigments common to members of the Haemodoraceae and in particular *Lachnanthes tinctoria* have been investigated in callus cultures. As the cultures age, the ratio of 2,5,6-trihydroxy-9-phenyl-phenalenone (*111*) (haemacorin aglycone) changes and it has been implied that metabolic demethylation of haemacorin aglycone takes place [153] (Fig. 21).

4 Terpenes

4.1 Monoterpenes

In a recent review on the biosynthesis of monoterpenes in plants, particular emphasis has been placed on studies involving the enzymology of biosynthesis [154]. It has been pointed out that biosynthetic studies on the formation of lower terpenoids in tissue cultures are few and that the study of carboxylases, for

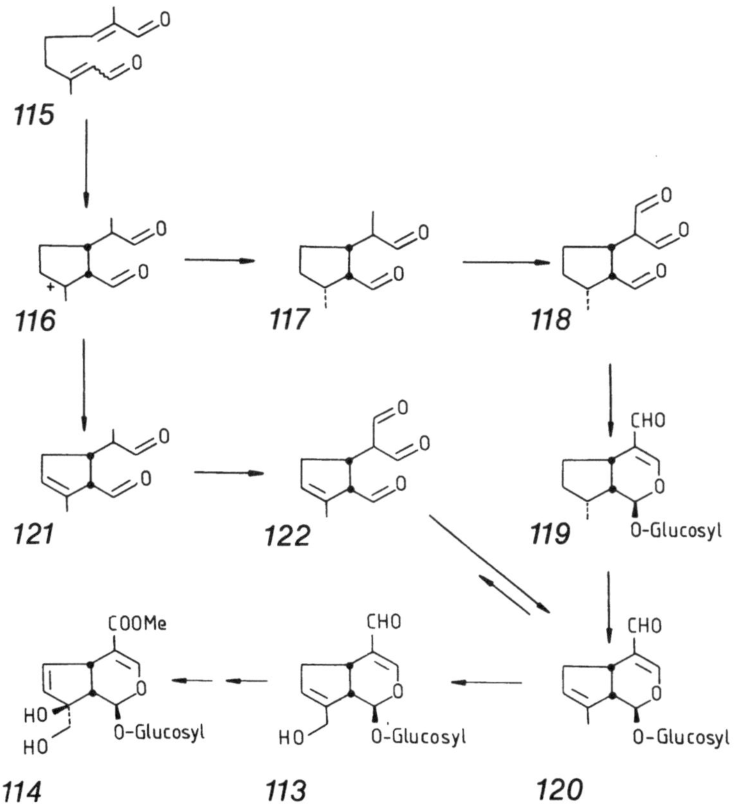

Fig. 22. Biosynthesis of tarennoside (113) and gardenoside (114) in cell suspension cultures of *Gardenia jasminoides*

example, obtained from tissue cultures could offer a breakthrough in our understanding of monoterpene biosynthesis.

Cell suspensions of the muscat grape (*Vitis vinifera*) have converted neral and geranial to the corresponding monoterpene alcohols, nerol and geraniol and the latter was further esterified [155]. It has been demonstrated, through the administration of various combinations of ^{13}C-labelled acyclic monoterpenes to *Gardenia jasminoides* f. *grandiflora* cell suspension cultures, that tarennoside (*113*) and gardenoside (*114*) are biosynthesized through cyclisation of 2E- or 2Z-10-oxocitral (*115*) to the iridodial cation (*116*) with subsequent randomization of the carbon atoms C-3 and C-11 (Fig. 22) [156,157]. However, the pathway after iridodial cation formation remained to be established. This problem has now been dealt with by a series of experiments in which ^{3}H- or ^{13}C-labelled monoterpenes were administered to these *Gardenia* cell suspension cultures. The experiments demonstrated that iridoid glycosides are biosynthesized after iridodial cation formation from 10-oxocitral (*115*) via 8-epiiridodial (*117*), 8-epiiridotrial (*118*), 8-epiiridotrial glucoside (boschnaloside) (*119*) and dehydroiridotrial glucoside (*120*). It is also possible that another route of biosynthesis to (*120*) is via dehydroiridodial (*121*) and dehydroiridotrial (*122*) (Fig. 22).

A number of investigations of biotransformations have been made recently with *Mentha* cell cultures. In one study, all cell lines investigated were able to convert (—)-menthone (*123*) to (+)-neomenthol (*124*) but none of these cell lines reduced (+)-isomenthone (*125*) to the corresponding alcohol [158].

Cell lines from different *Mentha* chemotypes were either capable or not capable of converting pulegone (*126*) into isomenthone (*125*). These results indicate stereospecifity in respect to both precursor and product in these plant cell lines. In order to determine whether only natural secondary metabolites or also other compounds which are similar to pulegone (*126*) were biosynthesized by *Mentha* cell suspensions,

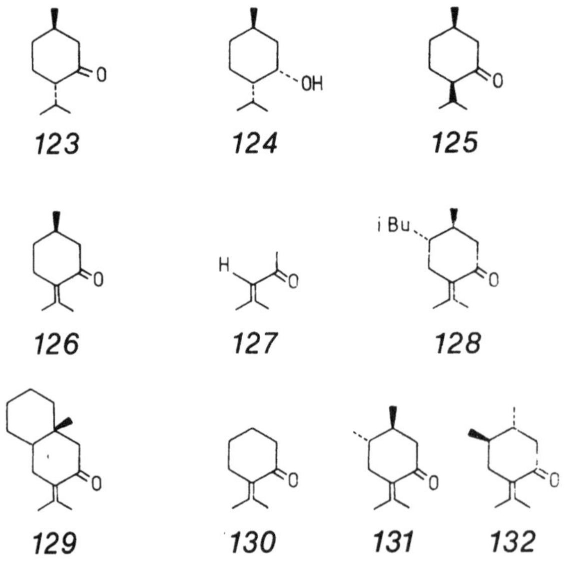

Fig. 23. Monoterpenes which have been examined as biotransformation products in *Mentha* cell suspension cultures

incubations with five unsaturated α,β-ketones were made. No conversion was detected with mesityl oxide (*127*), *trans*-6-*t*-butylpulegone (*128*) or 3-isopropylidine-9-methyl-decalone-2 (*129*). However, saturation of the α, β double bond of 2-isopropylidine cyclohexanone (*130*) and both isomers of *trans*-6-methyl pulegone (*131*, *132*) was observed in suspensions of those cell lines which were capable of pulegone transformation [159] (Fig. 23). Further to these experiments the same group investigated the effect of immobilization of these plant cells on enzyme activities [160]. Plant cells were entrapped by mixing suspended *Mentha* cells with either linear water soluble polyacrylamide-hydrazide chains followed by the stochiometric addition of glyoxal as the cross linking agent (PAAH-G entrapment), or in calcium alginate beads. The entrapped cells were found to be as efficient as the free cells in converting (−)-menthone (*123*) to (+)-neomenthol (*124*) and (+)-pulegone (*126*) to (+)-isomenthone (*125*). The successful entrapment of cells make possible the potential development of continuous biotransformation processes [160].

Cell suspension cultures of *Nicotiana tabacum* have been used to study the biotransformation of foreign terpenes such as α-terpineol (*133*), *trans*-β-terpineol (*134*), *trans*-β-terpinyl acetate (*135*) and linalool (*136*). The cultured cells have the ability to hydroxylate the C—C double bond and the allylic carbon atoms [161–164].

A series of carvone (*137*) isomers have also been fed to *N. tabacum* cell suspension cultures and it has been shown that the cells reduced, regio- and stereo-selectively, the C—C double bond adjacent to the carbonyl group [165]. Similar experiments with 1-acetoxy-*p*-menth-4(8)-ene (*138*) have revealed that the suspension cultures selectively hydroxylate stereospecifically the exocyclic double bond from the side opposite to the acetoxy group and regioselectively the allylic position of the exocyclic double bond [166] (Fig. 24).

Suspension cultures of *Cannabis sativa* are able to convert cannabidiol (CBD) (*139*) to both stereoisomers of cannabielsoin (CBE) (*140*) under normal growth conditions and hence it is possible that such a biotransformation could occur in the

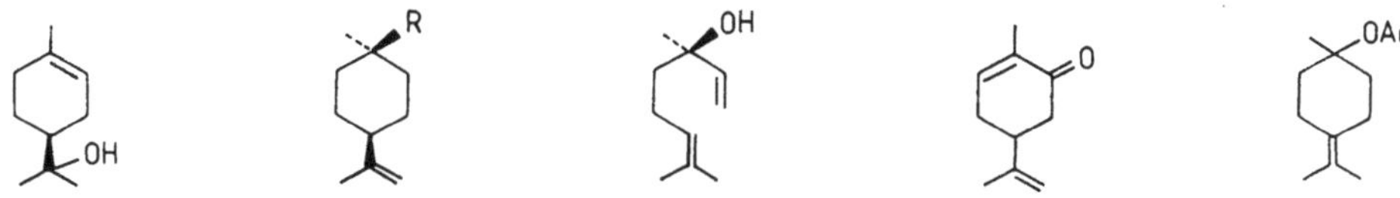

133 — 134 R=H, 135 R=Ac — 136 — 137 — 138

Fig. 24. Terpenes fed to *Nicotiana tabacum* cell cultures during biotransformation investigations

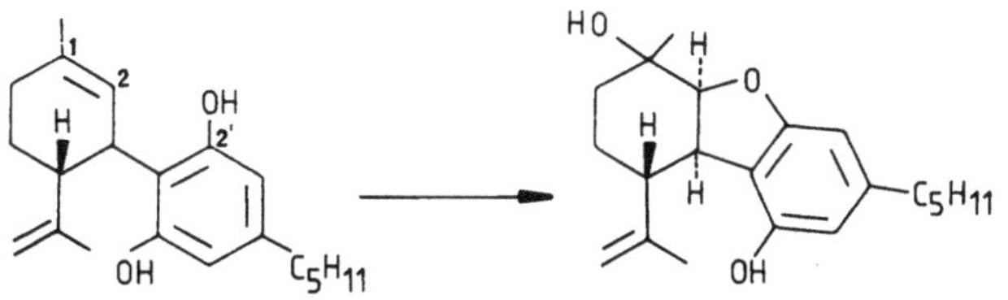

139 — 140

Fig 25. Conversion of cannabidiol (139) to cannabielsoin (140) by suspension cultures of *Cannabis sativa*

intact plant. It is possible that the mechanism of this conversion involves an epoxidation of the 1,2-double bond of CBD, followed by nucleophilic attack of the protonated epoxide at C-2 by the 2′-hydroxyl group [167,168] (Fig. 25).

4.2 Sesquiterpenes

In recent studies with tissue cultures of *Andrographis paniculata*, the incorporation of [2-^{13}C] and [3-^{13}C] leucines into paniculide A (*141*) and paniculide B (*142*) was investigated (Fig. 26). It was found that (3S)-3-hydroxy-3-methylglutaryl CoA (HMG-CoA) derived from leucine breakdown, is not incorporated intact into the paniculides A and B but that leucine is incorporated by breakdown to acetyl CoA and is subsequently incorporated via HMG-CoA and mevalonic acid [169].

Cell cultures of the liverwort *Calypogeia granulata* have recently been shown to synthesize dihydroazulene (*143*). The terpenoid origin of this compound has been investigated in biosynthetic studies employing ^{13}C-labelled acetate and difference ^{13}C NMR techniques. The biosynthetic route leading to 3,7-dimethylindene-5-carboxaldehyde (*144*) has also been clarified by ^{13}C NMR studies. This compound proved to be a trinorsesquiterpene which had undergone a skeletal rearrangement [170] (Fig. 26).

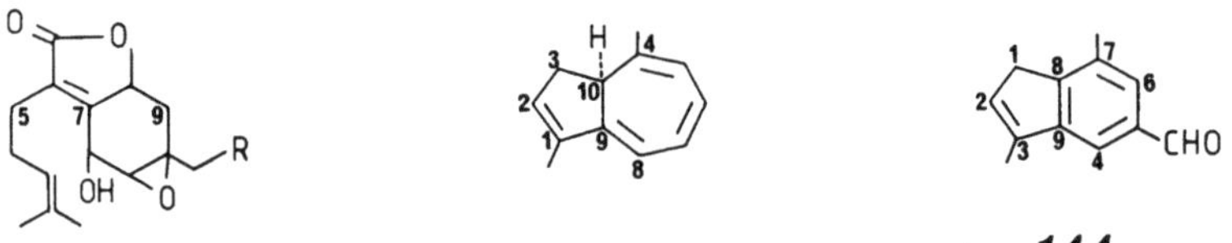

141 R = H **143** **144**

142 R = OH

Fig. 26. Sesquiterpenes investigated by plant cell culture techniques

4.3 Steroids

The metabolism of steroids in plant cell cultures [171] and the general biosynthesis of steroids in the Solanaceae [102] have been reviewed recently. The latter review covers the simple sterols, steroidal sapogenins, the steroidal glycosides and alkaloids, dealing with their metabolism as well as biosynthesis. To date, the extent to which plant cell cultures have been utilized in this area is small. Cell-free systems derived from tobacco cell cultures have been shown to convert acetate to squalene and also mevalonic acid to farnesol and squalene. The expoxidation of squalene to 2,3-oxidosqualene was also observed in cultures of tobacco cells incubated with radioactive acetate [102].

Tissue cultures of *Isodon japonicus* have been used to investigate the biosynthesis of oleanene and ursene-type triterpenes from [14-^{13}C] mevalonolactone and [1,2$^{13}C_2$] acetate [172]. Oleanolic (*145*), maslinic (*146*), 3-epimaslinic (*147*), ursolic (*148*) and 2α-hydroxyursolic (*149*) acids were isolated as their methyl ethers from callus cultures. All ^{13}C NMR signals of the olean-12-enes and urs-12-enes were assigned by chemical shift comparisons with those of a number of derivatives and by various NMR techniques. The ^{13}C-labelling patterns were elucidated with the ^{13}C-spectra of these triterpenes enriched with [4-^{13}C] mevalonolactone and [1,2 $^{13}C_2$] acetate. During the formation of the D- and E-ring systems, two rearrangements of the carbon skeleton for olean-12-enes and three rearrangements, including methyl migration from C-20 to C-19, for urs-12-enes were verified [172]. These results were entirely in accord with the biogenetic isoprene rule proposed previously for cyclisation of squalene to β-amyrin and α-amyrin and excluded an alternative mechanism proposed for α-amyrin biosynthesis, including a 19,19-dimethyl intermediate. Moreover, the C-23 and C-30 methyl groups were derived from the C-2 of mevalonate and the C-24 and C-29 methyl groups were derived from C-6 of mevalonate in both types of triterpenes. (3S)-2,3-oxido-squalene was confirmed as a precursor of 3α-hydroxytriterpenes as well as 3β-hydroxytriterpenes (Fig. 27).

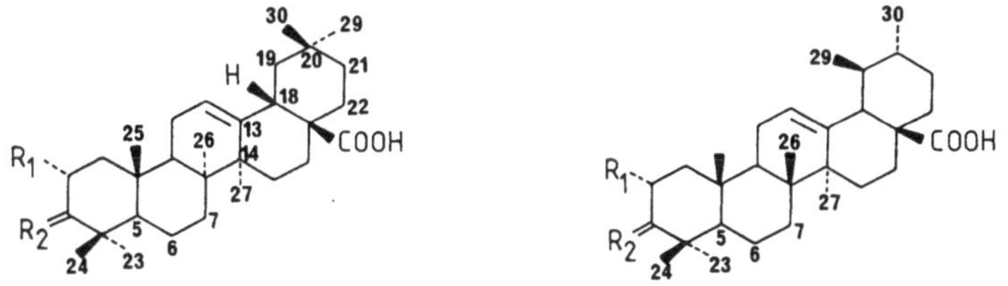

145 $R_1 = H$, $R_2 = $ OH/H

146 $R_1 = OH$, $R_2 = $ OH/H

147 $R_1 = OH$, $R_2 = $ H/OH

148 $R_1 = H$, $R_2 = $ OH/H

149 $R_1 = OH$, $R_2 = $ OH/H

Fig. 27. Biosynthesis of oleanene- and ursene-type triterpenes studied by the use of cell cultures of *Isodon japonicus*

Suspension cultures of brambles (*Rubus fructicosus*) normally contain Δ^5-sterols which can be suppressed by supplementing the medium with 15-azasterol. The 15-azasterol probably inhibited reduction of the Δ^{14}-bond and enabled the production of new cell lines [173]. *Cucurbita maxima* cell cultures have been used to investigate the biosynthesis of sterols and triterpenes using [2-^{14}C] acetate [174]. Synthesis of the cucurbitacins does not occur in callus tissues whereas there is a wide variety of 4,4-dimethylsterols indicating the high level of squaleneoxide cyclase activity. The feeding experiments enabled studies to be made on the biosynthesis of the sterol side chain which operates differently in the Curcubitaceae than in other higher plant families.

Cell cultures of *Agave wightii* have been found to be a good source of sapogenins and they produce gitogenin, hecogenin and tigogenin in both static and submerged cultures. Yields of papogenin were considerably increased by the addition of cholesterol to the media and overall yields of about 2% total sapogenin were obtained [175]. Similar increases in the yield of diosgenin have been observed when cholesterol was fed to *Dioscorea deltoidea* [176].

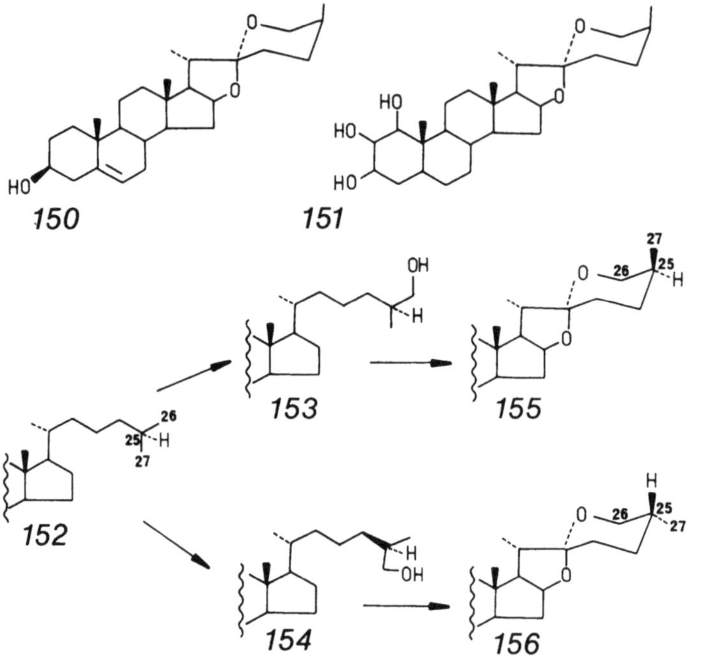

Fig. 28. Biosynthesis of steroidal triterpenes in cell cultures of *Dioscorea tokoro*

The biosynthesis of the (25S)-steroidal sapogenins, yamogenin (*50*), neotokorogenin (*151*) and their (25R)-isomers has been investigated using [1,2 $^{13}C_2$] acetate fed to *Dioscorea tokora* cell cultures [177]. The results indicated that the hydrogen atom at C-25 was introduced into the $^{24}\Delta$ double bond of intermediates predominantly from the 25-*si* face. The (25S)- and (25R)-sapogenins (*155*) and (*156*), respectively, were

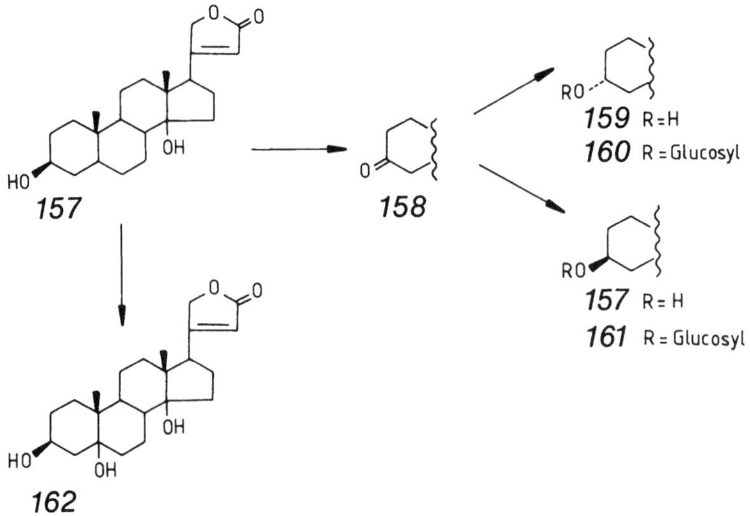

Fig. 29. Biotransformation of digitoxigenin (157) by cell suspension cultures of *Digitalis purpurea*

formed as a result of the oxidation of the *pro*-R and *pro*-S methyl groups of cholesterol (partial, *152*) via the 26-hydroxycholesterol isomers (partial *153* and *154*) (Fig. 28).

Biotransformations of *Digitalis* cardenolides continue to be of interest [178–180]. It has been shown that *D. purpurea* and *D. lanata* cell cultures oxidise digitoxigenin (*157*) at C-3 to form digitoxigenone (partial *158*) [181] whereas digoxigenin was converted to 12β-hydroxygenin [182].

Digitoxigenin (*157*) has been converted to periplogenin (*162*) by cell suspension cultures of *Daucus carota* and similarly, gitoxigenin and oleandrigenin have been converted to their 5β-hydroxyderivatives [180]. Further investigation into the biotransformation of digitoxigenin (*157*) by cell suspension cultures of *D. purpurea* have shown that oxidation to digitoxigenone (partial *158*) is followed by reduction to epidigitoxigenin (partial *159*) with final glucosylation of the hydroxyl groups with C-3 hydroxyl groups having α and β configurations (partial *160*, *161* respectively). Furthermore, it was confirmed that digitoxigenin was hydroxylated to periplogenin (*162*) (Fig. 29) [179].

5 Conclusions

It is obvious that since 1979 there has been considerable interest in the use of plant cell cultures for biosynthetic studies. It is interesting to note that despite the drawbacks which are entailed with the use of cultures, in particular uncertainty as to whether the biosynthetic pathway will be the same as in the differentiated plant, the technique has provided an abundance of information on biosynthetic intermediates, enzymes involved and the regulation of secondary metabolism. Indeed, in some instances, notably for *C. roseus*, the use of plant cell culture techniques has resulted in a more comprehensive understanding than has been reached previously by conventional in vivo techniques.

A major area of study has proved to be the biosynthetic pathways of indole alkaloids within cell cultures of *C. roseus*. It is ironic, that despite the considerable effort which has gone into the production and understanding of the biosynthesis of the dimeric anticancer alkaloids, vinblastine and vincristine, it has proved difficult to produce these alkaloids, de novo, in cultures whereas novel dimeric indole alkaloids have been produced by a species of the related apocynaceous plant, *Voacanga* [31]. The genera *Rauwolfia*, *Tabernaemontana*, and *Voacanga* are currently being investigated and the enzymes responsible for individual steps in the biosynthetic sequence of the contained alkaloids, are being isolated and characterized.

Although the study of indole alkaloids has had pride of place, considerable attention has now been devoted to other groups of alkaloids and in particular it is in the fields of isoquinoline alkaloid and quinolizidine alkaloid biosynthesis that the most progress has been made. One study resulting from work utilizing cell suspension cultures appears to be extremely significant and that is the reported presence of genes responsible for the production of quinolizidine alkaloids being widely distributed in plants which do not necessarily produce quinolizidines or any alkaloids whatsoever [70].

Plant cell cultures, particularly of parsley, have proved invaluable for determining the biosynthetic sequences within flavonoids and it has been possible to determine the characteristics of the enzymes involved in many of the stages of biosynthesis. It is to be anticipated that similar studies will continue, utilizing other species and investigating the biosynthesis of other phenolic constituents.

Apart from the steroids and the natural interest in *Dioscorea* for the production of diosgenin, little effort appears to have been made, since 1979, into the biosynthesis of monoterpenes, diterpenes and sequiterpenes. One previous reviewer has commented that the use of plant cell cultures might well give a breakthrough allowing for a more thorough understanding of the biosynthesis of monoterpenes [154].

Recent developments in sensitive analytical techniques, particularly of radioimmunoassay and enzyme assay have enabled biosynthetic studies within plant cell cultures to proceed at a faster rate than heretofore. The more widespread application of these sentitive techniques coupled with the use of plant cell cultures will ensure that a knowledge of the fundamental biosynthetic processes of plant secondary metabolism will continue to make rapid progress. The potential commercial production of useful compounds by cell culture techniques depends upon detailed investigations into the biosynthetic sequences which are involved [183]. The secondary metabolites shikonin and berberine, are now produced commercially by plant cell culture techniques and it is to be expected that other secondary products will be produced by these methods for the pharmaceutical, agricultural, chemical and food industries.

6 Acknowledgements

We are grateful to Mrs. L. Lisgarten for computer assisted literature retrieval, and to Mr. B. C. Homeyer for help in compilation of the manuscript.

7 References

1. Butcher, D. N.: In: Appl. and Fundamental Aspects of Plant Cell, Tissue and Organ Culture. (Reinert, R., Bajaj, Y. P. S., eds.), p. 668, Berlin, Heidelberg, New York: Springer 1977
2. Overton, K. H., Picken, D. J.: In: Progr. in the Chemistry of Organic Natural Products, (Herz, W., Grisebach, H., Kirby, G. W., eds.), Vol. 34, p. 249. Vienna, New York: Springer 1977
3. Kurz, W. G. W., Constabel, F.: In: Adv. Applied Microbiology, (Perlman, D., ed.) p. 209. New York, London: Academic Press 1979
4. Böhm, H.: In: Internat. Rev. Cytology, Supplement 11 B. (Vasil, I. K., ed.) p. 183. New York, London: Academic Press 1980
5. Dougal, D. K.: In: The Biochemistry of Plants Vol. 7 Secondary Plant Products, (Conn, E. E., and Stumpf, P. K. ed.) p. 21. New York, London: Academic Press 1981
6. Luckner, M., Nover, L., Böhm, H.: In: Secondary Metabolism and Cell Differentiation, Berlin, Heidelberg, New York: Springer 1977
7. Stöckigt, J.: In: Indole and Biogenetically Related Alkaloids. (Phillipson, J. D., Zenk, M. H., eds.) p. 113. London: Academic Press 1980
8. Herbert, R. B., Simpson, T. J. (senior reporters): Biosynthesis, Vol. 7. A Specialist Periodical Report. London, Royal Soc. Chem. 1983
9. Barz, W., et al.: Planta Med. *40*, 137 (1980)

10. Sasse, F., Heckenberg, U., Berlin, J.: Z. Pflanzenphysiol. *105*, 315 (1982)
11. Schallenberg, J., Meyer, E.: Planta Med. *42*, 133 (1981)
12. Pfitzner, U., Zenk, M. H.: ibid. *46*, 10 (1982)
13. Balsevich, J., Constabel, F., Kurz, W. G. W.: ibid. *44*, 231 (1982)
14. Balsevich, J., Kurz, W. G. W.: ibid. *49*, 79 (1983)
15. Spitzberg, V., Coscia, C. J., Krueger, R. J.: Plant Cell Reports *1*, 43 (1981)
16. Stöckigt, J., Soll, H. J.: Planta Med., *40*, 22 (1980)
17. Scott, A. I., et al.: Phytochemistry *19*, 488 (1980)
18. Kurz, W. G. W. et al.: ibid. *19*, 2583 (1980)
19. Kutney, J. P., et al.: ibid. *19*, 2589 (1980)
20. Kutney, J. P., et al.: J. Nat. Prod. *44*, 536 (1981)
21. Kohl, W., Witte, B., Höfle, G.: Z. Naturstoff. *36B*, 1153 (1981)
22. Guéritte, F., Langlois, N., Pétiard, V.: J. Nat. Prod. *46*, 144 (1983)
23. Kurz, W. G. W. et al.: Planta Med. *42*, 22 (1981)
24. Kutney, J. P., et al.: Helv. Chim. Acta *65*, 2088 (1982)
25. Baxter, R. L., et al.: J. Chem. Soc. Chem. Commun., 257 (1979)
26. McLauchlan, W. R. et al.: Tetrahedron *39*, 3777 (1983)
27. Kutney, J. P., et al.: Helv. Chim. Acta *64*, 1837 (1981)
28. Kutney, J. P., et al.: Heterocycles *16*, 1169 (1981)
29. Kutney, J. P., et al.: Helv. Chim. Acta *65*, 1271 (1982)
30. Kutney, J. P., et al.: Tetrahedron **39**, 3781 (1983)
31. Stöckigt, J., et al.: Helv. Chim. Acta *66*, 2525 (1983)
32. Pfitzner, A., Stöckigt, J.: Phytochemistry *21*, 1585 (1982)
33. Stöckigt, J., et al.: Planta Med. *39*, 73 (1980)
34. Zenk, M. H.: J. Nat. Prod. *43*, 438 (1980)
35. Stöckigt, J., et al.: Biochemistry *22*, 3448 (1983)
36. Kohl, W., Witte, B., Höfle, G.: Planta Med. *47*, 177 (1983)
37. Westekemper, P., et al.: ibid. *39*, 24 (1980)
38. Neumann, D., et al.: ibid. *48*, 20 (1983)
39. Courtois, D., Kurkdjian, A., Guern, J.: Plant Science Letters *18*, 85 (1980)
40. Renaudin, J.-P.: ibid. *22*, 59 (1981)
41. Renaudin, J.-P., Guern, J.: Physiol. Veg. *20*, 533 (1982)
42. Courtois, D., Guern, J.: Plant Science Letters *17*, 473 (1980)
43. Knobloch, K.-H., Bast, G., Berlin, J.: Phytochemistry *21*, 591 (1982)
44. Knobloch, K.-H., Berlin, J.: Z. Naturforsch. *35C*, 551 (1980)
45. Balagué, C., Wilson, G.: Physiol. Veg. *20*, 515 (1982)
46. Knobloch, K.-H., Hansen, B., Berlin, J.: Z. Naturforsch. *36C*, 40 (1981)
47. Lee, S.-L., Cheng, K.-D., Scott, A. I.: Phytochemistry *20*, 1841 (1981)
48. Roustan, J.-P., Ambid, C., Fallot, J.: Physiol. Veg. *20*, 523 (1982)
49. Stöckigt, J., Pfitzner, A., Firl, J.: Plant Cell Reports *1*, 36 (1981)
50. Pfitzner, A., Stöckigt, J.: Planta Med. *48*, 221 (1983)
51. Pfitzner, A., Stöckigt, J.: J. Chem. Soc., Chem. Commun. 459 (1983)
52. Pfitzner, A., Stöckigt, J.: Tetrahedron Letts. *24*, 1695 (1983)
53. Pfitzner, A., Krausch, B., Stöckigt, J.: Characteristics of vellosimine reductase, a specific enzyme involved in the biosynthesis of the *Rauwolfia* alkaloid sarpagine. Submitted for publication 1984
54. Pfitzner, A., Stöckigt, J.: Tetrahedron Letts. *24*, 5197 (1983)
55. Stöckigt, J., Pfitzner, A., Keller, P. J.: ibid. *24*, 2485 (1983)
56. Pawelka, K.-H., Stöckigt, J.: Plant Cell Reports *2*, 105 (1983)
57. Stöckigt, J., et al.: Z. Naturforsch. *37C*, 857 (1982)
58. Obato-Sasamoto, H., Komamine, A., Saito, K.: ibid. *36C*, 921 (1981)
59. Saito, K., et al.: Phytochemistry *21*, 474 (1982)
60. Rueffer, M., et al.: Febs. Lett. *129*, 5 (1981)
61. Schumacher, H.-M., et al.: Planta Med. *48*, 212 (1983)
62. Rueffer, M., Nagakura, N., Zenk, M. H.: ibid. *49*, 131 (1983)
63. Rueffer, M., Nagakura, N. and Zenk, M. H.: ibid. *49*, 196 (1983)
64. Kelleher, W. J., et al.: ibid. *40*, 127 (1980)

65. Rueffer, M., et al.: Tet. Lett. *24*, 2643 (1983)
66. Steffens, P., Nagakura N. and Zenk, M. H.: Planta Med. (in press 1984)
67. Amann, M., Nagakura, N. and Zenk, M. H.: Tet. Lett., (in press 1984)
68. Hsu, A.-F.: J. Nat. Prod. *44*, 408 (1981)
69. Tam., W. H. J., et al.: Phytochemistry *21*, 253 (1982)
70. Wink, M., Witte, L.: FEBS Letts. *159*, 196 (1983)
71. Wink, M., Hartmann, T., Witte, L.: Planta Med. *40*, 31 (1980)
72. Wink, M., Hartmann, T.: ibid. *40*, 149 (1980)
73. Wink, M., et al.: ibid. *38*, 238 (1980)
74. Wink, M., et al.: ibid. *44*, 15 (1982)
75. Wink, M., Hartmann, T.: J. Nat. Prod. *44*, 14 (1981)
76. Wink, M., et al.: Planta Med. *48*, 253 (1983)
77. Wink, M., Hartmann, T.: FEBS Letts. *101*, 343 (1979)
78. Wink, M., Hartmann, T.: Z. Naturforsch. *35c*, 93 (1980)
79. Schoofs, G., et al.: Phytochemistry *22*, 65 (1983)
80. Yamada, Y., Hashimoto, T.: Plant Cell Reports *1*, 101 (1982)
81. Mitra, G. C.: In: Plant tissue culture, genetic manipulation and somatic hybridisation of plant cells. (Rao, P. S., Heble, M. R., Chadha, M. S., eds.) p. 380, Proc. National Symp., Bhabha Atomic Research Centre, Bombay, India 1980
82. Koul, S., Ahuja, A., Grewal, S.: Planta Med. *47*, 11 (1983)
83. Griffin, W. J.: Naturwissenschaften *66*, 58 (1979)
84. Romeike, A.: Biochem. Physiol. Pflanzen *168*, 87 (1975)
85. Hiraoka, N., Tabata, M.: Phytochemistry *22*, 409 (1983)
86. Koelen, K. J., Gross, G. G.: Planta Med. *44*, 227 (1982)
87. Leete, E.: In: Alkaloids, chemical and biological perspectives. (Pelletier, S. W., ed.) Vol. 1, p. 85 New York: J. Wiley (1983)
88. Miller, R. D., Collins, G. B., Davis, D. L.: Crop Science *23*, 561 (1983)
89. Ravishankar, G. A., Mehta, A. R.: In: Plant tissue culture genetic manipulation and somatic hybridisation of plant cells. (Rao, P. S., Heble, M. R., Chadha, S., eds.) p. 342, Proc. of National Symposium at Bhabha Atomic Research Centre, Bombay, India 1980
90. Ohta, S., Yatazawa, M.: Biochem. Physiol. Pflanzen *175*, 382 (1980)
91. Berlin, J.: Phytochemistry *20*, 53 (1981)
92. Knobloch, K.-H., Beutnagel, G., Berlin, J.: Planta *153*, 582 (1981)
93. Knobloch, K.-H., Berlin, J.: Planta Med. *42*, 167 (1981)
94. Berlin, J., Kukoschke, K. G., Knobloch, K.-H.: ibid *42*, 173 (1981)
95. Knobloch, K.-H.: Plant Cell Reports *1*, 128 (1982)
96. Berlin, J., Knobloch, K.-H., Höfle, G., Witte, L.: J. Nat. Prod. *45*, 83 (1982)
97. Berlin, J., Witte, L.: ibid *45*, 88 (1982)
98. Baumert, A., et al.: Plant Cell Reports *1*, 168 (1982)
99. Baumert, A., Hieke, M., Gröger, D.: Planta Med. *48*, 258 (1983)
100. Zschunke, A., Baumert, A., Gröger, D.: J. Chem. Soc., Chem. Commun., 1263 (1982)
101. Osman, S. F., Zacharius, R. M., Naglak, D.: Phytochemistry *19*, 2599 (1980)
102. Heftmann, E.: ibid. *22*, 1843 (1983)
103. Frischknecht, P. M., Baumann, T. W.: Planta Med. *40*, 245 (1980)
104. Waller, G. R., Suzuki, T., Roberts, M. F.: Colloque Sci. Int. Café *9*, 627 (1980)
105. Waller, G. R., MacVean, C. D., Suzuki, T.: Plant Cell Reports *2*, 109 (1983)
106. Hirose, F., Ashihara, H.: Z. Naturforsch. *38c*, 375 (1983)
107. Hahlbrock, K., Grisebach, H.: Ann. Rev. Plant Physiol. *30*, 105 (1979)
108. Ebel, J., Hahlbrock, K. In: The Flavonoids: Advances in Research. (Harborne, J. B., Mabry, T. J. eds.) p. 641, London, New York: Chapman and Hall 1982
109. Noé, N., Seitz, H. U.: Z. Naturforsch. *38*c, 408, (1983)
110. Macheix, J. J., Suen, R., Ibrahim, R. K.: Biochem. Physiol. Pflanzen *176*, 195 (1981)
111. Koumba-Koumba, D., Macheix, J. J.: Physiol. Veg. *20*, 137 (1982)
112. Ellis, B. E.: Phytochemistry *22*, 1941 (1983)
113. Mizukami, H., et al.: ibid. *22*, 679 (1983)
114. Hahlbrock, K., Schröder, J., Vieregge, J.: In: Advances in Biochemical Engineering, Vol. 18

Plant Cell Cultures II. (Fiechter, A. ed.) p. 39, Berlin, Heidelberg, New York: Springer 1980
115. Wiermann, R.: In: The Biochemistry of Plants. Vol. 7. (Conn, E. E., Stumpf, P. K., eds.) p. 85 New York: Academic Press 1981
116. Duell-Pfaff, N., Wellmann, E.: Planta *156*, 213 (1982)
117. Wellmann, E., Schopfer, P.: Plant Physiol. *55*, 822 (1975)
118. Poulton, J.: In: The Biochemistry of Plants (Conn, E. E., Stumpf, P. K., eds), Vol. 7, p. 667. London, New York: Academic Press 1981
119. Egin-Bühler, B., Loyal, R., Ebel, J.: Arch. Biochem. Biophys. *203*, 90 (1980)
120. Light, R. J., Hahlbrock, K.: Z. Naturforsch. *35c*, 717 (1980)
121. Hahlbrock, K., Grisebach, H.: In: The Flavonoids. (Harborne, J. B., Mabry, T. J., Mabry, H., eds.) p. 866. London, New York: Chapman and Hall 1975
122. Heller, W., Hahlbrock, K.: Arch. Biochem. Biophys. *200*, 617 (1980)
123. Heller, W., Gardiner, S. E., Hahlbrock, K.: Hoppe-Seyler's Z. Physiol. Chem. *361*, 265 (1980)
124. Dixon, R. A., Dey, P. M., Whitehead, I. M.: Biochemica et Biophysica Acta *715*, 25 (1982)
125. Kreuzaler, F., et al.: Proc. Nat. Acad. Sci. USA, *80*, 2591 (1983)
126. Britsch, L., Heller, W., Grisebach, H.: Z. Naturforsch. *36c*, 742 (1981)
127. Gardiner, S. E., et al.: J. Biol. Chem., *255*, 10752 (1980)
128. Matern, U., et al.: Arch. Biochem. Biophys. *208*, 233 (1981)
129. Matern, U., Feser, C., Hammer, D.: ibid. *226*, 206 (1983)
130. Ayabe, S., et al.: Phytochemistry *19*, 2179 (1980)
131. Ayabe, S., et al.: ibid. *19*, 2331 (1980)
132. Ayabe, S., Furuya, T.: Tet. Lett. *22*, 2097 (1981)
133. Ayabe, S., Furuya, T.: J. Chem. Soc. Perkin Trans. I 2725 (1982)
134. Leweke, B., Forkmann, G.: Plant Cell Reports *1*, 98 (1982)
135. Stafford, H. A., Lester, H. H.: Plant Physiol. *68*, 1035 (1981)
136. Stafford, H. A., Shimamoto, M., Lester, H. H.: ibid. *69*, 1055 (1982)
137. Zähringer, U., et al.: FEBS Lett. *101*, 90 (1979)
138. Zähringer, U., Schaller, E., Grisebach, H.: Z. Naturforsch. *36c*, 234 (1981)
139. Kodama, T., Yamakawa, T., Minoda, Y.: Agric. Biol. Chem. *44*, 2387 (1980)
140. Bauch, H.-J., Leistner, E.: Planta Med. *33*, 105 (1978a)
141. Bauch, H.-J., Leistner, E.: ibid. *33*, 124 (1978b)
142. Inoue, K., et al.: J. Chem. Soc. Chem. Commun. 957 (1979)
143. Inoue, K., et al.: Phytochemistry *23*, 307 (1984)
144. Inoue, K., et al.: Tet. Lett. *21*, 621 (1980)
145. Inoue, K., et al.: J. Chem. Soc. Perkin Trans. *1*, 1246 (1981)
146. Fukui, H., et al.: Phytochemistry *22*, 453 (1983)
147. Fowler, M. W.: Nature *307*, 504 (1984)
148. Inoue, K.: J. Chem. Soc. Chem. Commun. 993 (1982)
149. Inoue, K., et al.: Phytochemistry *23*, 313 (1984)
150. Van den Berg, A. J. J., Labadie, R. P.: Planta Med. *41*, 169 (1981)
151. Yagi, A., Shoyama, Y., Nichioka, I.: Phytochemistry *22*, 1483 (1983)
152. Heide, L., Leistner, E.: ibid. *22*, 659 (1983)
153. Beecher, C. W. W., Sarg, T. M., Edwards, J. M.: J. Nat. Prod. *46*, 932 (1983)
154. Cori, M. O.: Phytochemistry *22*, 331 (1983)
155. Ambid, C., Moisseeff, M., Fallot, J.: Plant Cell Reports *1*, 91 (1982)
156. Uesato, S., et al.: Tetrahedron Lett⁻, *25*, 573 (1984)
157. Uesato, S., et al.: Chem. Pharm. Bull. *31*, 4185 (1983)
158. Aviv, D., et al.: Planta Med. *42*, 236 (1981)
159. Aviv, D., et al.: ibid. *47*, 7 (1983)
160. Galun, E., et al.: ibid. *49*, 9 (1983)
161. Suga, T., et al.: Chemistry Letts, 229 (1980)
162. Suga, T., Lee, Y. S., Hirata, T.: Bull. Chem. Soc. Jpn. *56*, 784 (1983)
163. Hirata, T., et al.: ibid. *54*, 3572 (1981)
164. Hirata, T., Lee, Y. S., Suga, T.: Chemistry Letts. 671 (1982)

165. Hirata, T., et al.: Phytochemistry *21*, 2209 (1982)
166. Lee, Y. S., Hirata, T., Suga, T.: J. Chem. Soc. Perkin Trans. I, 2475 (1983)
167. Hartsel, S. C., Loh, W. H.-T., Robertson, L. W.: Planta Med. *48*, 17 (1983)
168. Loh, W. H.-T., Hartsel, S. C., Robertson, L. W.: Z. Pflanzenphysiol. *111*, 395 (1983)
169. Anastasis, P., et al.: J. Chem. Soc. Chem. Commun. 1189 (1983)
170. Takeda, R., Katoh, K.: J. Am. Chem. Soc. *105*, 4056 (1983)
171. Stohs, S. J.: In: Advances in Biochemical Engineering, Vol. 16, Plant Cell Cultures I. (Fiechter, A., ed.) p. 85. Berlin, Heidelberg, New York: Springer 1980
172. Seo, S., Tomita, Y., Tori, K.: J. Am. Chem. Soc. *103*, 2075 (1981)
173. Schmitt, P., Scheid, F., Benveniste, P.: Phytochemistry *80*, 525 (1980)
174. Caputo, O., et al.: Planta Med. *49*, 176 (1983)
175. Sharma, O. P., Khanna, P.: J. Nat. Prod. *43*, 459 (1980)
176. Chowdhury, A. R., Chaturvedi, H. C.: Current Science *49*, 237 (1980)
177. Seo, S., et al.: J. Chem. Soc., Chem. Commun. 895 (1981)
178. Reinard, E., Alfermann, A. W.: In: Advances in Biochemical Engineering, Vol. 16, Plant Cell Cultures, I. (Fiechter, A., ed.) p. 49. Berlin, Heidelberg, New York: Springer 1980
179. Hirotani, M., Furuya, T.: Phytochemistry *19*, 531 (1980)
180. Jones, A., Veliky, I. A.: Planta Med. *42*, 160 (1981)
181. Stohs, S. J., Rosenberg, H.: Lloydia *38*, 181 (1975)
182. Alfermann, A. W., et al.: In: Plant Tissue Culture and its Bio-Technological Application. (Barz, W., Reinhard, E., Zenk, M. H., eds.) p. 125. Berlin, Heidelberg, New York: Springer 1977
183. Fowler, M. W.: In: Plant Biotechnology. (Mantell, S. H., Smith, H., eds.) p. 3. London, New York: Cambridge University Press 1983

Immobilized Plant Cells

A. Rosevear and C. A. Lambe
A.E.R.E., Harwell, Oxfordshire,
OX11 ORA, United Kingdom

Immobilization of plant cells is still a young technology but this review illustrates the potential it has for improving the production of phytochemicals. The similarity between the immobilized state and whole tissue is emphasized and the way in which this can be used to advantage is discussed. The delicate nature of plant cells limits the methods which can be used for their immobilization but once bound, the cells can be used for both biotransformations and perhaps more importantly biosynthesis. The slow metabolism and peculiar nutrient demands of plant cells restrict the reactor configurations which can be considered for long term operation. However preliminary estimates of production costs suggest that immobilized cells would be more economic than cultured cells for the production of extracellular secondary metabolites.

1 Introduction

Differentiated cells of higher plants have an awesome ability to synthesize a diverse range of natural products, many of which form the basis of the existing agricultural, forestry and pharmaceutical industries. Recent developments in tissue culture techniques have made it possible to use less differentiated plant cells to produce some of the more valuable phytochemicals under controlled conditions [1], offering the prospects of higher productivity and greater flexibility in production of secondary metabolites.

However, by removing the cells from the organized structure of the whole plant, tissue culture inevitably places greater demands on the environmental conditions within a reactor than is the case with agriculture. The benefits of rapid growth in a controlled reactor are balanced by penalties which stem from the need to aseptically supply the cells with nutrients and to remove metabolic products. These restrictions are often manifested as limits to cell density in the reactor (150–200 g wet wt per l), heterogeneous callus tissue, cellular damage due to fluid movements and viscosity problems due to polymeric products.

The first objective of the plant cell culturist has been to generate biomass and over the last 10 years the greatest success has been achieved by those who have treated plant cell culture as a rather specialized form of a bioprocess [1], in which mild agitation is used to give efficient mixing and an homogeneous cell suspension. However, the production of valuable secondary metabolites is frequently associated with slow growing or senescent cells, and Yeoman [2,3] has gone sor far as to suggest that secondary product formation is inversely related to growth. He argues that cell-cell interaction is the easiest method of promoting cyto-differentiation, the prerequisite of product synthesis in the intact plant and in many non-selected cell cultures [4].

Immobilization of plant cells, i.e. physically restraining the cells on a fixed support, can be seen as an intermediate state between homogeneous suspension culture and the highly structured tissue matrix of the whole plant. By deploying individual cells or small clumps in a protective and stable mechanical framework, a rudimentary degree of organization can be imposed on the system, giving control over mass transfer and over heterogeneity in the microenvironment of the cell. The technologist is able to take advantage of the controlled growth of cell biomass in suspension culture, and to benefit from the contact-induced cytodifferentiation previously only possible in callus or mature tissue.

2 Immobilized Biocatalysts

Immobilization is a generic term used to describe methods of restricting the free movement of any independent biocatalyst. Immobilized plant cells are a natural progression from the work on less complex biocatalysts and the current approach to plant cell immobilization often reflects lessons learnt in this earlier period. However, it should be remembered that in immobilizing enzymes [5], prokaryotic microorganisms [6], or even simple eukaryotes such as yeast [7], one is often transforming a naturally homogeneous biocatalyst to a heterogeneous form. In contrast, plant cells are genetically predisposed to associate and may survive for several years in a natural cooperative structure. Thus, in common with animal cells [8], carefully manipulated plant cells may present fewer problems to the immobilization technologist than to the process engineer who is to use the immobilized biomass. The production of high value phytochemicals from immobilized plant cells has proved a major stimulus to work in this area and the current review covers the literature from the first successful report by Brodelius et al. [9] to the middle of 1983.

Immobilization usually results in a change in the kinetics and process requirements of a biocatalyst. Even though cells are already insoluble particles, agglomeration on a subsidiary structure will usually result in their taking on the characteristic of a heterogeneous catalyst once immobilized. As yet no specific information has been tabulated for plant cells but the problems of mass transfer to concentrated biocatalytic particles, diffusion barriers around particles in a stream of fluid and microenvironment changes due to the build up of products are common features of the use of immobilized enzymes [10].

As in the case of enzymes, the loss of biocatalytic activity should be minimized during immobilization so that for cells in which the biosynthetic capacity is to be exploited, activity can be crudely interpreted as retention of cell viability. A major parameter against which an immobilized enzyme reactor is judged is operational half life, but since the cells often have the capability to replace lost enzymic activity or even dead cells (by replication), this concept is of less fundamental importance here. Long term operation of a cell reactor can best be achieved by a careful balance of process conditions to give a stable population of slow growing cells. This then avoids uncontrolled growth, which would otherwise lead to blockage of the reactor or appearance of free cells in the product.

3 Benefits of Plant Cell Immobilization

The value of biocatalyst immobilization has been forcefully illustrated by the industrial success of enzymes such as glucose isomerase [11]. Many of the benefits of immobilization are general, varying in importance depending on whether the biocatalyst is a single enzyme, a fast growing prokaryote or a large eukaryotic cell. The benefits can be classified as those resulting from, (a) retention of catalyst in the reactor: (b) high concentration of catalytic activity: (c) microenvironmental control: (d) quantitative and rapid removal of biocatalyst from the product, and (e) uncoupling of growth and production phases.

We believe that judged against these criteria, the benefit of an immobilized system over the competing free biocatalyst approach is at its greatest for slow growing eukaryotic cells such as plants. We shall take these points in turn.

a) Retention of biocatalyst ensures that its full productivity is exploited, i.e. it operates under optimum conditions for its full lifetime. This contrasts with the premature loss of catalyst which occurs in homogeneous systems. Thus, immobilization maximizes the use of capital invested in preparing the catalyst and reduces upstream costs. This advantage is obviously directly related to the cost of the catalyst, being greater for the more expensive biocatalyst, such as eukaryotes. The reduction in nutrient and medium requirements has additional benefits on the downstream side of the reactor, as removal of unwanted biomass is unnecessary the product recovery is simplified.

b) The residence time needed to achieve a particular reaction is directly related to the concentration of biocatalyst in the reactor. In free cell systems cell density must often be limited by restrictions on mass transfer and the build up of inhibitors, the problem being most acute with eukaryotes. By deploying the cells in a spacially fixed system, fluid can flow rapidly past the densely packed biocatalyst giving efficient mass transfer of nutrient and products. Besides the reduction in contact time and the benefits this gives in process response time, immobilized cell reactors will be more compact and will generate fewer of the by-products often associated with prolonged process time. This compactness can lead to dramatic reduction in capital costs both in the reactor and in downstream purification [12].

c) One of the major concerns in using a free biocatalyst is the maintenance of homogeneity within the reactor by efficient mixing. However, eukaryotic cells, derived from complex organisms, often respond to local metabolite gradients and such heterogeneity may be important in the expression of secondary metabolism and product transport. Immobilization can be used to impose a particular microenvironment on a cell or to allow neighbouring cells to influence one another through short range chemical mediators. The control of dissolved gases is particularly important in plant cell culture and Fowler [13] has noted that even simple gassing of free cell cultures can cause shear damage, while attempts to mechanically stir cultures can cause severe lysis. The mechanical protection offered by immobilization is particularly advantageous for plant cells in this respect and offers more freedom to the process engineer in matching gas flow to cellular needs. Thus the effects of oxygenation and stirring can be decoupled.

d) The absense of suspended particles with neutral buoyancy obviously simplifies downstream processing and product recovery from biological fluids, as it eliminates a separaration step. Post-reactor modification of the product is minimized when quantitative retention of the catalytically active material is achieved and deactivation of the catalyst in the presence of product is unnecessary. The tendency of many cells to turn over products makes immobilization the method of choice for recovering labile components or intermediate metabolites.

e) The uncoupling of growth and production phases is probably the most important benefit of cell immobilization. Since the biomass is retained in the reactor its continuous replenishment is not required. Thus, cells can be grown under optimized conditions in a single separate operation and then used in a continuous reactor at dilution rates well in excess of those which would cause wash-out in free cell systems. The resulting high fluid velocity permissible even with these slow growing cells not only increases the efficiency of mass transfer, supplying fresh nutrients, but more importantly sweeps away inhibitory products, creating a favourable concentration gradient across the cell membrane.

Thus immobilization of plant cells offers substantial processing advantages, justifying the effort of defining the complex operational needs of these cells.

4 Unique Features of Plant Cell Immobilization

The sensitivity of plant cells to environmental factors and their slow rate of metabolism raise particular problems in using them in an immobilized form. Cell death during immobilization and vulnerability of the cell reactors to infection by fast growing bacteria can present serious difficulties. However, these same characteristics can be turned to advantage, for instance by the use of hormones for cellular control and the exploitation of their potential longevity which seems to be a corollary to the slow growth of these large cells.

Some of the experience with immobilization of live microbial cells [6] can be usefully applied to work on plant cells but the greater complexity, size and fragility of plant cells makes these the most difficult multienzyme systems to which immobilization technology has yet been seriously applied. In common with other eukaryotes, these cells show localization of specific functions in specialized subcellular organelles (e.g. nuclei, mitochondria, chloroplasts and storage vacuoles). The part played by light in tissue cultured cells is unclear [14], the questions of not only light intensity but also photoperiod adding extra uncertainties to work with plants. Unlike microbial cells, freshly synthesized products seem to accumulate in storage vacuoles, presenting particular problems if continuous reactor operation is intended. The enormous size of plant cells compared to prokaryotes (up to 10^4 times greater by volume) makes them vulnerable to mechanical shear and osmotic shock during immobilization, though the generally low rate of metabolism and lower oxygen demand [1], other consequences of their greater size, may be an advantage during the short period of stress involved in many immobilization methods.

A major characteristic of higher plants is that individual cells within a particular cell line are capable of diverse expression of their basic genetic information. This differentiation can dramatically alter both the physical and chemical characteristics of a particular cell in a way not found with other biocatalysts. Such changes are thought to be under the control of various plant hormones (auxins, cytokinins and gibberellins), some of which may be added in the media while others are synthesized directly by the cells, thus affecting cells in their immediate environment. This response to external non-nutrient stimuli reflects the origins of plant cells in cooperative association with other specialized cells, and while it offers greater flexibility of response compared to prokaryotes, the very complexity of the system makes it difficult to understand, predict or control at our present state of knowledge. Consequently the full potential of the immobilized state may not be realized as quickly as simple culture techniques.

The lack of linear response to known hormones and the unpredictable effect of external agents as diverse as light, hydrocarbons, gas balance, temperature and pH as well as nutrient levels makes control of plant cells more difficult than that of prokaryotes particularly in the specific microenvironment around the immobilized cell. Even without mutation, semi-permanent, epigenetic changes can occur in cell lines resulting in marked changes in secondary metabolism or growth. Thus until stable, regulated expression of the plant genotypes can be better controlled, free enzymes or prokaryotic cell systems will be preferred except where the uniqueness and high value of the eukaryotic cell product justifies the risk of immobilization and prolonged use. This will almost certainly limit interest to secondary metabolites, and if the earlier observations on differentiation are accepted, to slow growing cells in close contact with one another.

5 Products from Immobilized Plant Cells

The above discussion makes it clear in our view, that although plant cells are capable of performing almost any bioconversion, their use is not justified for bulk production of primary metabolites or the synthesis of simple secondary metabolites for which chemical or enzymic methods are possible. However, plant cells are of great interest for the production of high value secondary products currently derived from agriculture [15]. For many of these products, immobilization will provide the most attractive route, although for those in which biomass is a vehicle or diluent of the active component (e.g. food flavour, aroma or colour), mass cell culture may be preferable. Furthermore, political considerations, small markets and the accumulated agricultural experience of many centuries is likely to make flavour and aromas the least amenable targets for new technology of any type.

Figure 1 contrasts the main characteristics of product formation by mass cell culture, agriculture and an immobilized system. By deploying cells at high density and

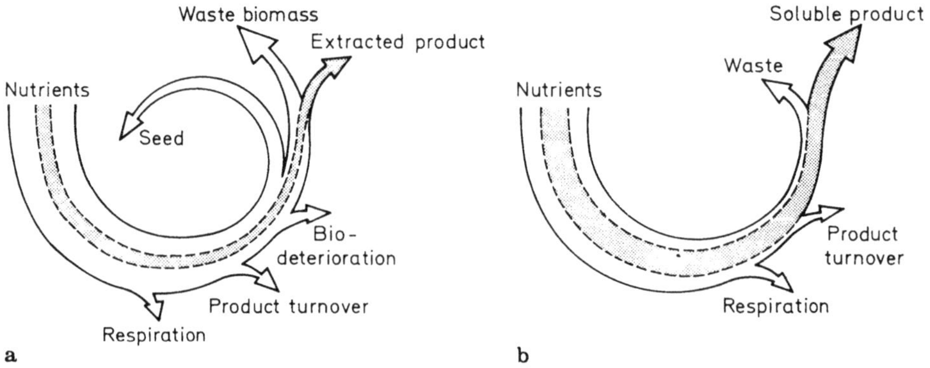

Fig. 1 a and b. Characteristics of main technologies for exploitation of eukaryotic cells; **a.** Agriculture or mass cell culture; **b.** Immobilized cells

limiting cell growth it will be possible to divert maximum metabolic activity to product formation rather than wasteful biomass. This is obviously most useful when the desired product is a minor component of normal biomass, but it also presupposes that the product is excreted by the cells without cellular damage and that it is soluble in the fluid flowing through the reactor. Many gallenicals and food additives however, fall into this category, although the former condition was for some time thought to be a serious limitation. There is now growing evidence that many important products can be found in the fluid around metabolizing cells [16,17,18,19].

Our excitement at the prospects for immobilized cells does not depend solely on their use as a substitute for agriculture or forestry. The novel environment created by cell immobilization provides opportunities previously not open to the natural product chemist. The synthesis of new chemical analogues is already possible [20], and could be used to widen the range of products available particularly if new genetic information is also introduced into the cell. The prospect of bringing down the cost of interesting phytochemicals such as hormones and lectins could open up new markets for plant products as agrichemicals or process aids. Phytoallexins [21], natural biocides produced in response to infection, could be produced by treatment of immobilized cells with non-infective extracts in a way not possible in the open field or growing cell cultures.

By analogy with immobilized enzymes, success in such new technology is more likely with totally new products, for example anti-cancer agents for which no established crops or plantations exist [22]. Obviously the more complex the product the more likely it is that plant cells would be preferred. Thus, glycosylation and other post-translational modifications may be limited to eukaryotes and the larger the number of asymmetric centres in a molecule, the less likely it is that chemical synthesis could compete.

The identification of the "right" product is probably a more serious limitation on the exploitation of plant cells than the technical problems identified above. The choice of a suitable cell line must precede any discussion of the relative merits of free cell culture or immobilization. Although we feel that for biochemical synthesis, immobilized systems are more likely to be competitive, it should be remembered that the biomass for immobilization must come from some form of mass culture and that advances in this area [1], are equally important for the eventual success of this approach.

6 Immobilization

6.1 Supply of Cells

To facilitate mass transfer within the immobilized cell matrix, the tissue used for immobilization should be substantially homogeneous and finely divided. Undifferentiated cells from suspension culture [13,23], provide the best source of such a consistent biomass and most immobilization methods have routinely used these. Cells are grown prior to immobilization in a dedicated culture vessel, usually a shake flask at current levels of technology, before being removed by coarse filtration and rapidly immobilized under aseptic conditions. Although this handling must cause some stress

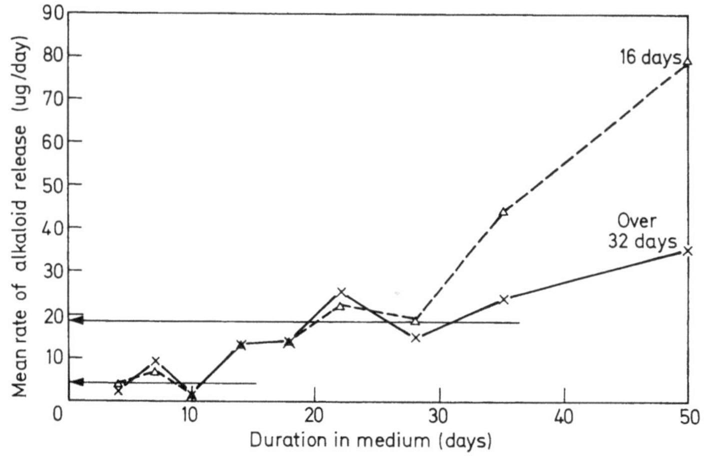

Fig. 2. The time cells spent in induction medium prior to immobilization can affect the level of subsequent alkaloid production and cell longevity — data from *Catharanthus* cells in alginate/acrylamide [24]

to the cells by disrupting normal metabolism, there is little immediate impact on viability perhaps because of their generally slow metabolism and protective cell wall.

Since cell growth after immobilization is discouraged it is important that the cells used are predisposed to produce the required metabolites. Yeoman [3,4], has observed that secondary products are normally best synthesized in slow growing differentiated cells. Thus rapid pre-growth in suspension culture, despite its more efficient biomass generation compared to callus culture, may seem at odds with the aims of this technology. However, the key features of differentiation can be induced by physical changes in the environment of the cells, for example by proximity [24], and so immobilization per se can induce expression of the desired metabolism from cells grown in homogeneous suspension. This effect can be enhanced by prior growth in an induction medium which promotes secondary metabolite formation, and we have found that the period of pre-conditioning can have a marked effect on both productivity and longevity of the immobilized cells (Fig. 2). This phenomenon is probably related to the near cessation of further cell division once all of the cells have been immobilized.

Once immobilized, the cells are no longer subject to the mechanical shear which keeps suspension cultures homogeneous. Consequently, the slow cell growth manifests itself as small calli adhering to the immobilization matrix [17]. The high packing density of the immobilized cells and the resultant cell/cell contact achieved confers on them many of the characteristics of differentiated/callus tissue. However, their distribution in a controlled matrix ensures that the fluid movement is similar to that in a homogeneous suspension, with the added advantage that the system can operate at high dilution rates without loss of biomass.

When immobilized, the cells can only exhibit properties which are already inherent within them: immobilization merely provides a convenient method of exploiting these properties. Furthermore, although the immobilization process may seriously stress the cell, subsequent productivity and longevity are more likely to be features of the cell itself rather than the immobilization method used. The latter simply affects the number of viable cells and the ease with which they can be manipulated and is of little consequence if the chosen cell line selected is unstable or genetically unsuitable for high productivity.

Methods of immobilization can be divided into those which depend on gel entrapment and on surface binding. In the former, the cells are held within a porous structure and products diffuse freely through this, while in the latter the cells are bound asymmetrically to a surface by chemical or physical means and products diffuse directly into the medium.

6.2 Gel Entrapment

Gel methods have been the preferred approach for the immobilization of viable cells, including plant cells. Since the gel structure is formed around the cell, there is, in principle, no chemical modification of the cell wall, although some graft polymers may be formed by in situ polymerization [25]. The cell thus remains suspended in a normal fluid environment bounded usually by a hydrogel structure fairly similar to that found in whole tissue. This arrangement will often be capable of accomodating limited cell expansion and imposes some orientation on the cell while not seriously restricting free movement of solution within the fluid.

Calcium alginate gel, a polysaccharide derived from seaweed and cross-linked by the divalent cation, has proved to be very useful for immobilization of delicate cells [26]. The first reported immobilization of plant cells [9], employed this method to give beads containing *Morinda, Digitalis* and *Catharanthus* cells. The gentleness of the technique was illustrated by the high retention of cell viability throughout the 22 day period in nutrient medium.

The technique is not only mild and reproducible, but requires only simple equipment. The cells are suspended in a solution of sodium alginate which is then extruded through a defined orifice using either a pump or dropping pipette. The resulting droplets are gelled on contact with a solution of calcium chloride (50–100 mM), which can be made up in nutrient medium to minimize shock to the cells. The viscosity of the alginate limits the practical range of polymer to concentrations of 2–3 %, although it should be noted that autoclave sterilization permanently reduces the viscosity of alginate solutions. The alginate gels can be formed with many di- and polyvalent cations and a number of microbial cells have been immobilized in aluminium [27], and barium alginate. The toxicity of some metal ions causes problems with plant cells and since calcium is a constituent of nutrient media it will inevitably replace any other ion in time and so is routinely chosen for most purposes.

An outer skin of gel forms around the droplet immediately upon entering the salt solution giving a clean, well defined bead. However, complete gelation takes several minutes, particularly with the dilute solutions being used to minimize osmotic shock. The cells are evenly distributed in the gel and can be recovered in a viable form by dissolving the alginate in citrate, or other calcium chelating agents such as EDTA and polyphosphate.

The technique does though, have two main limitations. The first is that of controlling droplet size. The viscosity of the cell/alginate solution can result in large droplets (several millimetres in diameter) being formed, independent of the orifice size used, while attempts to employ very narrow jets can often lead to blockage by cell aggregates. The larger beads can suffer from mass transfer limitations and we have noted necrosis in the core of such beads with a concomitant increase of biomass on the periphery (I. Hislop, personal communication 1981). We have overcome this by forming the gel as a sheet reinforced by cotton cloth [28]. This composite can be made to any thickness and handled as a single unit.

The other major problem with this technique is associated with the metal ion used for cross linking. Calcium ions can be important triggers of many biological mechanisms [29], and are known to affect cell/cell adhesion in plant systems [30]. However, since the strength of the gel varies with the amount of divalent metal present it is difficult to vary one independently of the other in an alginate complex. Supplementation of the calcium levels in the medium is usually ineffective for prolonged, continuous use of the immobilized cells and the gel is often replaced by invasive tissue growth which alters mass transfer and may fragment with time.

Phosphate in the medium is probably the main cause of decalcification and Jones and Veliky [31], have shown that this ion can be replaced by MES buffer. This provides additional buffering capacity but as phosphate is an essential nutrient its omission may prejudice the long term stability of the cell reactor. Knobloch and Berlin [32], have however found that secondary metabolism is enhanced at low phosphate levels, perhaps as a result of low growth [2], and so some flexibility in phosphate levels may be desirable. Despite these limitations, the calcium alginate gel remains the simplest and most popular method of immobilization, particularly for short term studies.

Brodelius and Nielsson [33] compared a number of gels for the immobilization of *Catharanthus* cells. They concluded that cell survival was very poor when low molecular weight monomers (e.g. acrylamide and glutaraldehyde) were used and only pre-formed polymers, particularly those of marine origin such as alginate and carrageenan, were sufficiently mild to give viable cells held in a strong gel.

However, all these gels depend on non-covalent bonds for their structure and are likely to have poor long term stability under continuous operation. Consequently, we have explored methods of forming more stable networks based on chemically cross-linked hydrogels. We have found that strong, open gels can be formed by polymerizing acrylamide in the presence of biopolymers such as xanthan gum or sodium alginate [28]. By reducing the concentration of monomer to 5% w/v, *Catharanthus*, *Nicotiana* and *Daucus* cells can be immobilized with over 90% retention of cell viability.. Grafting of the acrylamide to the biopolymer probably occurs while the latter provides more protection for the cells and its viscosity prevents sedimentation of the cells during the gelation. Cotton or stainless steel mesh is usually incorporated in the gel to provide tear strength and to simplify mechanical handling of the resulting gel sheet. The gels are cross-linked with this acrylamide and so do not require calcium for their stability, although those containing alginate do exhibit some changes in appearance if calcium levels are varied.

Felix and Mosbach [34] have increased the stability of permeabilized *Catharanthus* cells by cross-linking the agarose gels with hexamethylene diamine and glutaraldehyde. However, this does have a deleterious effect on key enzyme activities. They have successfully immobilized the same cells in a polyurethane foam block formed by mixing wet cells with a Hypol 3000 polymer from W. R. Grace. Glutaraldehyde hardening, as with agarose, reduced initial activity substantially but increased long term stability. Finally, *Mentha* cells have been immobilized in a cross-linked polyacrylamide-hydrazide gel and employed in the reductive biotransformation of monoterpenes.

Hollow fibre modules have been used successfully for growing animal cells and show some potential for retaining plant cells in a reactor system. The cells are held on the shell side of the unit while culture fluid is recirculated through the fibres. Low and medium molecular weight nutrients and cellular products pass freely through the intervening membrane while the cells are held between the fibres. Polyphenols have been produced in such systems using both soya bean [62] and carrot [63] cells over periods of up to 700 h.

6.3 Surface Binding

Non-specific binding of cells to surfaces such as glass is a common feature of cell suspension cultures, especially in conditions where fluid shear is low. However, this adhesion is difficult either to predict or control. Larkin [35] has introduced some degree of specificity in the type of adsorption by adsorbing protoplasts to a matrix made up of cross-linked albumin and the lectin Concanavalin A. Jirku et al.[36] have covalently bound *Solanum* cells to porous, high surface area, beads of polyphenylene oxide which had been pre-activated with glutaraldehyde. The absence of low molecular weight toxins and the mildness of the chemical reaction prevented the loss in viability which might be expected from direct reaction with cell wall components.

The natural tendency of suspension cultured cells to aggregate has been exploited by Lindsey et al. [18], who found that after introduction of particles of polyurethane foam into suspension, cultures of *Capsicum* or *Daucus*, the cells collected preferentially in the pores of the foam. The resulting aggregates were stable and could be used in reactor systems.

Rhodes et al. [37] have reported a similar phenomenon with hop and beetroot cells grown in the presence of nylon pan scrubbers. Initial evidence points to enmeshment of the growing cell mass with the interconnected pores of the matrix as being the major reason for success of these remarkably easy immobilization techniques. The process is simple and more rapid than simple growth of cells on the surface, i.e. about 2 days. However, it is still slow compared with gel immobilization and perhaps more importantly, is under less direct control. The density, distribution and disposition of the cells is ill defined and it is not yet clear how efficient mass transfer is within these clumps which are in effect calluses of a predetermined size.

A further example of how plant cells can be encouraged to adhere to surfaces is shown by Bornman and Zachrisson [38], who have used Cytodex microcarriers, originally developed for animal cell culture, to immobilize protoplasts of *Datura innoxia*. In the presence of Concanavalin A, loading reached 70% of the theoretical monolayer covering.

7 Use of Immobilized Plant Cells

Immobilization offers opportunities both for production of valuable industrial biochemicals and for studies of basic cellular biochemistry. The former is currently of greatest interest and will be discussed in terms of biotransformations (i.e. conversions of one complex molecule into a related product), and biosynthesis in which complex products are formed de novo from simple nutrients. The various examples are summarized in Table 1.

7.1 Biotransformations

This is in principle the most straightforward use of plant cells in which a limited number of cellular enzymes are used to convert an externally supplied compound, using the cell's primary metabolism merely to regenerate appropriate co-factors. In many respects it resembles the use of immobilized enzymes e.g. the use of strictosidine synthetase immobilized on Sepharose [39], and probably relies less on the metabolic condition of the cell than does de novo synthesis. Low nutrient levels and relatively high

Table 1. Use of immobilized plant cells

Cells	Immobilization method	Product	Operation	Ref.
BIOTRANSFORMATION				
Digitalis lanata	Alginate bead	Methyl digitoxin → methyl digoxin	60 d batch change 70 d continuous	[40] [41]
Daucus carota	Alginate bead	Digitoxigenin → periplogenin	12 d batch change 3 weeks	[31,42] [60]
BIOSYNTHESIS				
Beet	Nylon pad	Sucrose → β cyanin	A few days	[37]
Capsicum frutesans	Nylon net on polyurethane foam	Capsaicinoids	Prolonged period	[18,4]
Catharanthus roseus or *Vinca rosea*	Alginate beads	Tryptamine → ajmalicine	90 h batch	[9]
	Agar or gelatin	Tryptamine → ajmalicine	2 weeks batch	[33]
	Alginate bead	Sucrose → ajmalicine	21 d	[41,16]
	Alginate/acrylamide cloth	Sucrose → ajmalicine	220 d batch change	
	Xanthan/acrylamide cloth	Sucrose → Serpentine	180 d batch change	[28]
Nicotiana species	Xanthan/acrylamide cloth	Sucrose → alkaloid	Shake flask	[28]
	Alginate beads	Sucrose → nicotine	Air lift reactor	[19]
Solanum aviculare	Polyphenol oxide bead	Sucrose → steroid glycosides	Column 270 h	[36]

substrate concentrations can be tolerated, although the toxicity of some secondary products (e.g. menthols used in *Mentha* cultures by Alfermann and Aviv), can limit the operation range. However, the major limitation is likely to be the permeability of the cell both to substrate and product, since under normal circumstances the reactions involved would serve only to interconvert intracellular compounds.

Hydroxylation of cardiac glycosides has proved an interesting application of immobilized plant cells. Alfermann and Schuller[40], used *Digitalis lanata* cells in alginate beads in a process aimed at converting β-methyl digitoxin into the clinically more useful β-methyl digoxin. In a batch process, the fluid bathing the cells was changed every second day over a period of 59 days and resulted in a higher cell productivity than that of a free cell system. Although further use of these beads then became restricted by decalcification and loss of physical integrity, Brodelius et al.[41] successfully used a column of similar beads for up to 70 d. Using alginate immobilized cells of *Daucus carota* Jones and Veliky[42] achieved 5′ hydroxylation of digitoxigenin into periplogenin over a period of 12 d, but noted a deterioration in biocatalytic activity due to the accumulation of inhibitory levels of the toxic substrate in the cells.

Except for certain specialist biotransformations, it seems unlikely that true immobilized plant cells can compete effectively in this area with a less complex biocatalyst such as immobilized enzymes or microbial cells. The need to selectively deactivate unwanted side reactions and obtain 2-way membrane permeability to secondary metabolites presents formidable problems yet has limited applicability if the problem is solved.

7.2 Biosynthesis

In contrast, biosynthesis poses a different set of problems whose solution is likely to be more generally applicable to a range of unique products. However, it is inevitable that the more tortuous the biosynthetic pathway involved, the less efficient is the incorporation of carbon and nitrogen in the product. Furthermore, complex secondary metabolism is often incompletely understood and impossible to control with any degree of certainty. Thus, a semi-empirical approach must be adopted in the use of hormones and cell mediators to minimize side reactions and maximize product formation. The problem is exacerbated by the inherent genetic stability of diploid plant cells and the low incidence of abnormal, hyperproducing clones.

Though these are problems common to all work on plant cells, an additional requirement with immobilized cells is obtaining an extracellular product so that continuous operation is feasible. Thus, the report by Brodelius et al. [9] that secondary metabolism was stimulated in immobilized *Morinda* cells was encouraging but still left the major problem of how the anthraquinone dye could be removed continuously from the cells. This same paper reported the first attempt to elute product from fixed cells by externally modifying the permeability of alginate bound *Catharanthus* cells. When incubated in nutrient medium containing tryptamine these cells produced ajmalicine which accumulated in vacuoles. However, as a result of saturating the medium with chloroform, membrane permeability was changed and the alkaloid appeared as an extracellular product.

Although it is likely that in this case the extracellular product resulted from irreversible cell lysis, the experiment had led to a very productive line of research on selective permeabilization of plant cells. Felix et. al. [43] showed that dimethyl sulphoxide enhanced the transport of ajmalicine from storage vacuoles to the culture medium without affecting cell viability. The general permeabilization of the membranes made possible the introduction of isocitrate into the cells, thus increasing the supply of intracellular co-factors. Felix and Mosbach [34] have further shown that the stability of the permeabilized cells is increased by treatment with glutaraldehyde and hexamethylene diamine. Jones and Veliky [31] showed that cells could survive on medium containing chloroform at half saturation, a level sufficient for permeabilization.

The concentration of solvent required for complete permeabilization varies with plant species [64], *Catharantus roseus* requiring only 5% DMSO while *Daucus carota* and *Datura innoxia* need 10% and 25% respectively. On the basis of these observations a model production system was set up in which *C. roseus* cells immobilized in alginate or agarose were subjected to alternating periods of growth and product extraction during which ajmalicine stored in the cell was released.

However, removal of membrane components by solvent is a rather drastic step and is likely to debilitate individual cells due to loss of key metabolites along with the secondary metabolites. We have found that with *Catharanthus* and *Nicotiana*, there is no need to resort to such harsh treatment in order to promote release of products from immobilized cells. Unlike free *Catharanthus* cells, alginate/acrylamide bound cells were found to liberate substantial amounts of ajmalicine and serpentine into the medium if this was changed regularly in batch operation [16]. This steady accumulation of alkaloid in the spent medium has been maintained for up to 220 d following twice weekly changes of the cells to fresh, hormone-free medium. The cumulative amount released exceeded both that present in the original cells and the amount that these cells would have produced if left in the original nutrient medium (Fig. 3). Cell viability remained around 50% for much of this experiment, and

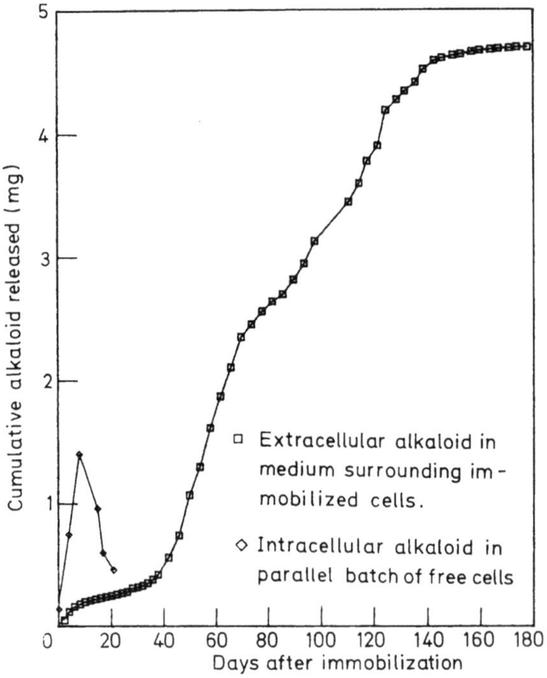

Fig. 3. Long term productivity of *Catharanthus* cells immobilized in alginate/acrylamide [16]

although some callus-like cell growth did occur, we believe that lysis and cryptic growth was insufficient to account for this level of release.

The condition of the cells prior to immobilization is important in determining the long term stability and productivity of immobilized *Catharanthus* [24]. Figure 2 shows that free cells taken from an induction medium before metabolism were switched to alkaloid synthesis (around day 10), never produced very much alkaloid after immobilization. Cells immobilized in late exponential or early stationary phases continued to release product at high levels for a prolonged period while very old cells rapidly released their alkaloid and died. These observations are consistent with a system in which the catalytic performance is determined by those cells which survive immobilization and in which subsequent recolonization under constant operating conditions is minimal.

Brodelius et al.[41] reported stimulation of ajmalicine synthesis in alginate immobilized *Catharanthus* cells and found that after 21 d, 80% of this alkaloid was in the fluid medium. Using *Digitalis lanata* cells in alginate beads, Alfermann[17], achieved de novo synthesis of cardiac glycosides which accumulated in the medium when this was regularly changed over a period of 170 d[18].

Thus, evidence is growing to show that long term production of extracellular, secondary metabolites is possible using viable, immobilized plant cells. As yet the mechanism is unclear but viewed in engineering terms, the phenomenon is of great practical importance even if it is merely due to lysis and re-growth. In *Catharanthus* we have observed the same effect for cells immobilized by several methods and additional circumstantial evidence (not shown here), supports the view that the release of a number of products into the fluid around immobilized cells is not a lytic process. Extracellular polymeric materials, including polyphenols, carbohydrates and a number of distinct, (though as yet unidentified), proteins are also released from the bound cells, in different proportions to the polymers released by free cells. Stimulation of extracellular production[44], and changes in cellular productivity[45], have been noted in immobilized prokaryotes and simple eukaryotes and such basic metabolic functions as respiration rate are known to increase in fixed microorganisms[46]. Mattiasson and Hahn-Hagerdal[47] have recently hypothesized that reductions in water activity and/or oxygen deficiency are responsible for the metabolic changes found with immobilized cells, including the apparent diversion of activity away from cell growth. Ellwood et al.[48] have invoked the chemiosmotic theory to interpret the changes in metabolism which occur when microorganisms become attached to surfaces. The asymmetry of the cellular microenvironment plays a major part in the explanation and the subtlety of this theory could prove very interesting when applied to immobilized cells in general.

Yeoman et al.[2,4] have invoked arguments involving increased cell-cell contact and gradients of intercellular messenger compounds to explain both increased secondary metabolism and also subsequent trans-membrane transport of these products. It is inevitable that the microenvironment of an immobilized cell will differ substantially from that of a similar cell in free suspension, particularly in the local concentrations of extracellular products in the diffusion layer around the cell. Plants are cooperative organisms and specialization of function must often demand transport of metabolites between cells. Muller et al.[49] showed that the alkaloid nicotine was synthesized in the roots of whole plants and transported to the leaves while Fairburn et al.[50] illustrated that opium alkaloid was transported in the plant within latex vesicles. Deus and Zenk[51] have shown that active transport mechanisms exist in tonoplast membranes of *Catharanthus* cells while Renalldin[52] claims that non-specific transport of ajmalicine across membranes can be achieved by changes in pH.

Thus, we believe that immobilization can and perhaps already has, provided the microenvironmental conditions to stimulate transport of secondary metabolites which are stored in vacuoles in free cells. Our own observations indicate that the fluorescent alkaloid serpentine still accumulates in the vacuoles of immobilized *Catharanthus* cells, but that the constant removal of the extracellular product appears to promote fresh synthesis, presumably by preventing feed-back inhibition or repression. Immobilization can thus be viewed as a convenient means of rapidly creating a microenvironment to mimic that in the productive parts of the whole plant. This not only slows growth and promotes diversion of nutrients to secondary products rather than unwanted biomass, but may also further promote product formation by maintaining a flux of products from the cells. To what extent the examples quoted above are typical can only be shown by further work on other cell lines, though initial studies are encouraging. However, until the natures of the "mediators", "messengers" and stimulatory conditions are better characterized, true control and predictable operation cannot be assured.

7.3 Non-synthetic Uses

Schnabl et al. [53] have illustrated that alginate immobilization stabilizes delicate photoplasts of *Vicia* and have suggested this as a means of transporting and storing protoplasts. Scheurick et al. [54] have proposed that protoplasts within a gel bead provide a convenient experimental system to study the effects of osmotic change, and Larkin [35] has suggested that immobilized plant cells are a useful tool for investigation of phenomena such as cell/cell interactions. The increased stability of protoplasts on immobilization leads to the suppression of ethane production and other signs of slow cell death [65]. This effect is reversed in the presence of toxins such as phenol or mercury and the amount of ethane production by alginate immobilized protoplasts has been suggested as a method of monitoring environmental poisons [66].

The sloughing off of growing cells, mentioned earlier as a problem during product formation has been turned to advantage by Morris et al. [67], who immobilized coarse cell aggregates or calluses and then collected the free cells which were shed from the gel beads during subsequent cell growth. This method was particularly useful for producing fine suspensions from fast growing cell lines but unfortunately, these lines reverted to their former appearance after a few sub-cultures.

The exploitation of the convenient mechanical properties of immobilized plant cells is only just beginning but these systems seem to offer many opportunities for the biochemist and physiologist to study basic cell metabolism and membrane phenomena in a fixed and isolated experimental system.

8 Reactors

Deploying immobilized cells in the substrate solution is as important as the actual method of immobilization. This presents more problems with living plant cells than simple enzyme systems. In the latter the substrate merely provides a feedstock for the reaction while in the whole cell reactor the fluid flow must also provide the essential nutrients required for cell viability and biocatalytic activity. Unsatisfactory fluid flow, whether as a result of poor fluid distribution or interruption of substrate supplies, can cause irreversible loss of reactor activity. Thus, although immobilization offers opportunities for manipulation of cellular productivity, it also makes greater demands on the control systems even for a simple process. As yet very little attention has been devoted to such studies, but we will attempt to identify some of the general considerations in the design of plant cell reactors.

Conditions for high catalytic utilization are often likely to be at odds with those necessary for maintaining cell viability for a number of reasons. In simple catalyst systems water activity is often kept low, substrate concentration high and non-reactive process aids at the lowest possible level. However, live cells are unlikely to survive prolonged periods under this kind of stress and particularly in the case of de novo synthesis, where the substrate is also a nutrient, complete conversion of the substrate will inevitably stress the cells. This is further complicated if high osmolarity is needed to promote secondary formation as suggested by Zenk [55].

The high volumetric flow of oxygen needed to supply the high density of cells in a fixed reactor, especially if the cells have a higher respiratory demand when immobilized, will inevitably cause inefficient back-mixing, require liquid/gas decoupling equipment and pose additional sterility problems.

Since long operation is essential if immobilization is to be justified, plant cell reactors will probably be run at lower substrate concentrations, to lower levels of conversion and with more process aids (e.g. non-metabolized additives), than is usual with simple enzyme systems. The back-mixing due to gassing and the long residence times needed to give product formation make single-pass, plug flow columns unattractive for plant cell reactors. A recycling reactor configuration with a continuously fed buffer tank (Fig. 4), can be operated at fluid velocities giving

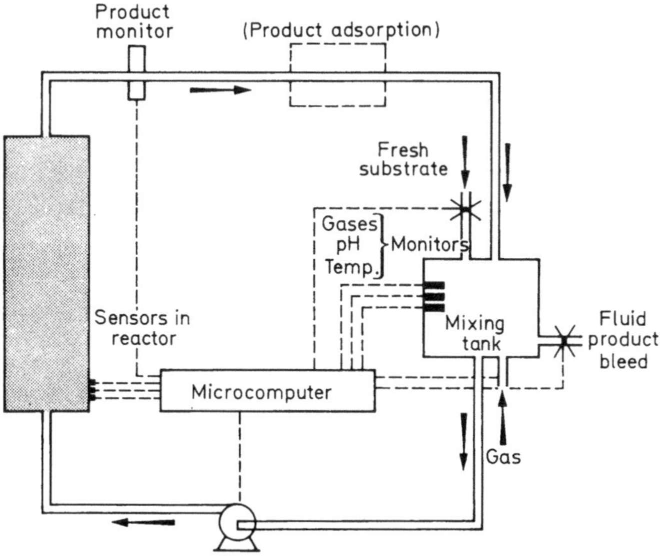

Fig. 4. Continuously operated recycle reactor suitable for immobilized plant cells

a long average residence time. The high superficial fluid velocity could reduce mass transfer resistance around the immobilized cells and maintain a high flux of nutrients and gases to the reactor while removing products rapidly from the presence of the cells, preventing both feed-back inhibition and product degradation. Such a system would limit nutrient starvation and product toxicity thus preventing loss of cell viability at the downstream end of the reactor, a phenomenon already noted with microbial reactors used to totally remove a nutrient from the fluid in a plug flow reactor [56]. This inevitably implies incomplete utilization of the substrate, adding to process costs. Adsorption of the product from the process stream could make recycling and makeup of the sterile spent liquor feasible, but would require additional improvements in downstream processing of plant products (e.g. high specificity, pseudo-affinity type adsorbents).

Since immobilized plant cells are mechanically robust compared with free cells, gas sparging need not be limited by concern over cellular damage nor need fluid velocities be limited by consideration of cell washout. Supply of nutrient is of greater concern in an immobilized cell reactor since cell densities could reach 70% of the volume compared with 15% in freely suspended cell culture. The control over mass transfer offered by immobilization is apparent when one considers that free cell densities of more than 30% by volume are almost impossible to stir and handle [55]. Although this represents an improved process intensity in the immobilized cell reactor, the large volume of intracellular water (90% of cell volume) means that enzyme densities cannot reach the level found in immobilized enzyme systems. Furthermore, the high cell density and the fixed position of immobilized cells make it difficult to define the light intensity to which such cells may be subject. Thus, it would be difficult to operate a cell reactor which required a specific level of illumination or photoperiod to give a high yield. Our limited experience suggests that in *Catharanthus* cells alkaloid synthesis proceeds in the dark. While there are many photosensitive products which might benefit by protection from light, secondary metabolites derived from other pathways, e.g. via phenylalanine ammonia lyase, may be limited.

Immobilized cells can be formed in either particles or sheets. Particles and beads are very versatile and the same type of material can be packed into columns or beds of different dimensions, simplifying scale-up. However, good flow distribution, important in the maintenance of cell viability, can be difficult to achieve. In down-flow reactors compaction can occur while in up-flow reactors, particularly if gas sparging is needed, buoyancy could restrict flow rates and cause channelling. Fluidized or expanded beds are likely to be more convenient for particulate materials although the reduced biocatalyst density in these reactors is a drawback. Sheets of immobilized cells provide an alternative method of deploying biocatalysts [57], and we have found that gel sheets of plant cells are more convenient to handle in the laboratory than are beads. However, sheets are rarely selfsupporting, need secondary structures within the reactor and must be custom made for each reactor design. On the other hand, they can be packed so as to provide a lumen of controlled dimensions, giving more defined fluid flow through the reactor. Coaxially wound sheets [57], baffled tanks, flat plate reactors [58], or modified plate and frame presses some which have been used for other immobilized biocatalysts, provide convenient starting points for reactor design using existing process plant.

In practice, most workers have so far limited themselves to simple reactors which are convenient for analytical sampling rather than high productivity. Shake flasks used as batch reactors [16], and simple packed bed reactors [40,33], have been used most frequently, but Mantell and Smith [19] have used alginate immobilized tobacco cells in an air lift reactor.

Veliky and Jones [60] have made preliminary studies on the problem of injecting air into a continuously operating column reactor. Direct injection at the base of the reactor gave better substrate conversion than external aeration in a separate limb, indicating that any large system may need to accommodate gas rising through the main reactor.

9 Reactor Operation

Absolute sterility of all equipment, nutrients and gases is essential if any plant cell reactor is to be kept free of bacterial and fungal contamination. Since most production media are likely to discourage plant cell growth, the systems are even more vulnerable to fast growing microorganisms with simple nutrient requirements. Aseptic handling is costly and inconvenient and so both immobilization and reactor design should be aimed at minimizing manipulation of materials. Once loaded, a reactor is only susceptible to infection via the nutrient fluid, but the long operational life needed to justify an immobilized cell system places absolute importance on substrate sterility.

Most current media include a diverse range of salts and vitamins in addition to the carbon source. This is not particularly costly (e.g. \$ 50 per m^3) compared with that for animal cells, but it is likely that several minor components could be omitted without loss of cell viability [15]. Even these apparently defined media can change unpredictably before use. For instance, we have found that sucrose is progressively inverted during autoclaving, and invertase produced by the cells can lead to increases in osmolarity by hydrolysis of the sucrose over periods of a few days. The poor buffering capacity of many traditional media is insufficient to cope with the metabolites produced by cultured cells or even the range of carbon dioxide levels encountered. As a consequence of this, such systems tend to operate at a pH dictated by cellular metabolism rather than that predetermined by the experimentalist.

Jirku et al. [36] have used sucrose alone as a feedstock while Knobloch and Berlin [32] and Yeoman et al. [4] have used simplified media. However, it is unclear how the longevity of immobilized cells is affected by these changes, particularly since we have observed that intracellular starch grains can give short term protection from nutrient starvation. It is likely that some simplification of media is possible, especially since most media used at present are designed to satisfy the needs of growing cells, rather than the low growth requirements of immobilized cells. However, there is evidence that high molarity stimulates secondary products [55], and it may be necessary to incorporate non-metabolized compounds such as sugar alcohols in the medium.

Goldstein et al. [59] have attempted a cost analysis for production of biochemicals from tissue cultured plant cells. Using existing free cell technology, they estimate a manufacturing cost of \$ 551 per kg at a production level of 10^4 kg per year. Of this almost 40 % is attributed to raw materials. By assuming 90 % recycle of biomass and a 10-fold increase in rate of product formation, the cost falls to \$ 324 per kg of which only 16 % is attributed to raw materials. This degree of recycle or biomass reuse, is expected as a result of immobilization. Furthermore, immobilization does not require the element of cost for biomass recovery included by Goldstein, while prolonged operation committed to secondary metabolism under low growth is likely to increase the rate of product formation. Thus, immobilized cells are likely to prove more cost effective than free cells at present levels of technology. The prerequisite for immobilized cell operation, an extracellular product, simplifies initial clarification and recovery of the product but does not prevent problems of recovery from dilute aqueous solution. This may increase the final recovery costs for some immobilized cell products, especially when the phytochemical could have been used in a semi-purified form (e.g. within the biomass). Nevertheless, for many valuable gallenicals and agrichemicals the benefit of an efficient reactor should be dominant and make immobilization competitive with agricultural methods.

Goldstein et al. [59] made further assumptions such as doubling of reactor biomass to 500 g wet weight per l, similar to levels which can be achieved by immobilization, and this brought projected costs to \$ 165 per kg of product. Their final assumptions of a 10-fold increase in cell productivity and large scale economies at a production rate of 10^6 kg per year are not yet justifiable, but could realistically reduce costs to \$ 18 per kg, well within the cost range for common plant products and bulk chemicals. Although these calculations are only estimates, it is interesting to note that the two main characteristics of cell immobilization, diversion of raw material away from biomass and increased biocatalyst concentration, contribute to a 70 % reduction in costs compared to a similar free cell system. Thus, it would appear that immobilization should be the preferred method of producing soluble phytochemicals, although it must be remembered that a source of mass cultured suspension cells is essential to supply the initial biomass for such reactors.

10 Conclusions

Plant cell immobilization is now becoming an established technique. At present, reactor systems and operational conditions for exploiting these materials are still in their infancy and the dearth of cell lines giving high yields of soluble products still limits the application of the technology. However, these do not present insuperable problems and the prospects for substantially reduced manufacturing costs using immobilized plant cells provide strong commercial incentives to demonstrate industrial scale operation of such systems. This could be done most easily for existing plant products but the longer term prospects for totally new groups of products which exploit the characteristics of slow growing cells under maximum containment and control offer even more exciting opportunities. Thus immobilized plant cells demand the interest of scientist, technologist and engineer as a challenging and potentially profitable topic of work.

11 Acknowledgement

The authors are currently involved in a Research Programme on the Immobilization of Plant and Animal Cells and are grateful to the Materials, Chemicals and Vehicles Requirements Board of the United Kingdom Department of Trade and Industry, for sponsorship of this work.

12 References

1. Fowler, M. W.: Chem. Ind. (7), 229 (1981)
2. Yeoman, M. M.: Plant Cell Culture Symp., London, Oyez 1982
3. Yeoman, M. M. et al.: Plant Cell Culture (Sala, Parisi, Cella and Ciferra eds.), Amsterdam: Elsevier 1980
4. Yeoman, M. M. et al.: Differentiation in Vitro (Yeoman and Truman, eds.), Cambridge: C.U.P. 1982
5. Trevan, M.: Immobilized Enzymes, New York: Wiley 1980
6. Cheetham, P. S. J.: Topics in Enzyme and Ferment. Biotech. *4*, 189 (1980)
7. Kolot, F. B.: Proc. Biochem. *15*, (7), 1 (1980)
8. Rosevear, A. et al.: Topics in Enzyme and Ferment. Biotechn. *7*, 13 (1983)
9. Brodelius, P. et al.: FEBS Lett. *103*, 93 (1979)
10. Engasser, J. M. et al.: Appl., Biochem. Bioeng. *1*, 128 (1976)
11. Hemingsen, S. H.: ibid. *2*, 157 (1979)
12. Daniels, M. J.: Food Process Eng. *2*, 103 (1980)
13. Fowler, M. W.: Prog. Ind. Microbiol. *16*, 207 (1982)
14. Seibert, M. et al.: Plant Tissue Culture as a source of biochemicals (Staba, E. J. ed.) p. 123, Baton Rouge: C.R.C. Press 1980
15. Staba, E. J.: Plant Tissue Culture as a source of biochemicals, Baton Rouge: C.R.C. Press 1980
16. Lambe, C. A. et al.: Enzyme Eng. *6*, 137 (1982)
17. Alferman, A. W.: Seminar Symp. Ser. Leicester, Reading, Soc. of Experimental Biology 1982
18. Lindsay, K. et al.: FEBS Lett. *155*, 143 (1983)
19. Mantell, S. H. et al.: Seminar Symp. Ser. Leicester, Reading, Society of Experimental Biology 1982
20. Reinhard, E. et al.: Adv. Biochem. Eng. *16*, 49 (1980)
21. Smith, D. A.: Biologist *28* (2), 69 (1981)
22. Humphreys, J.: New Scientist 513 (1982)
23. Kurz, W. G. W. et al.: Continuous Culture of Cells (Calcott P. H. ed.), p. 141. Baton Rouge: C.R.C. Press (1981)
24. Lambe, C. A. et al.: Europ. Pat. Appl. 82301571.4 (1982)
25. Somerville, H. J. et al.: Eur. J. Appl. Microbiol. *4* (2), 75 (1977)
26. Kierstan, M. et al.: Biotech. Bioeng. *19* (3), 387 (1977)
27. Hackel, U. et al.: Eur. J. Appl. Microbiol. *1*, 291 (1975)
28. Rosevear, A.: Europ. Pat. Appl. 8130400.1 (1981)
29. Hughes, et al.: Chem. Brit. *18* (3) 196 (1982)
30. Dougall, D. K.: Plant Tissue Culture as a source of biochemicals (Staba, E. J. ed.), p. 21, Baton Rouge: C.R.C. Press 1980
31. Jones, A. et al.: Eur. J. Appl. Microbiol. Biotech. *13* (2), 84 (1981)
32. Knobloch et al.: Z. Naturforsch. *35c*, 551 (1980)
33. Brodelius, P. et al.: FEBS Lett. *122* (2) 312 (1980)
34. Felix, H. R. et al.: Biotechnol. Lett. *4*, (3), 181 (1982)
35. Larkin, P.: Recent Adv. Phytochem. *15*, 135 (1981)
36. Jirku, V. et al.: Biotechnol Lett. *3* (8), 447 (1981)
37. Rhodes, M. J. C., et al.: Plant Cell Culture Symp., London Oyez 1982
38. Bornman, C. H. et al.: Plant Cell Rpts. *1*, 151 (1982)

39. Pfitzer, U., et al.: Planta Med. *46*, 10 (1982)
40. Alfermann, A. W., et al.: ibid. *46*, 10 (1982)
41. Brodelius, P. et al.: Europ. Pat. 224434 A2 (1981)
42. Jones, A. et al.: Can. K. Bot. *5* (11), 20954 (1981)
43. Felix, H. R.: et al.: Anal Biochem. *116*, (2), 462 (1981)
44. Yamamoto, K. et al.: Biotech. Bioeng. *16*, 1589 (1974)
45. Navarro, J. M. et al.: J. Appl. Microbiol. *4*, 243 (1977)
46. Marlipar, A. et al.: Biotech. Lett. *1*, 65 (1980)
47. Mattiasson, B. et al.: Eur. J. Appl. Microbiol. Biotech. *16*, 52 (1982)
48. Ellwood, D. C. et al.: Phil. Trans. R. Soc. Lond. B. *297*, 517 (1982)
49. Muller, E. et al.: Nova Acta Leopold. Suppl. *7*, 133 (1976)
50. Fairburn, J. W. et al.: Phytochem. *13*, 1133 (1974)
51. Deus, B. et al.: Biotech. Bioeng. *24* (9), 1065 (1982)
52. Renalldin, J. P.: Plant Sci. Lett. *22* (9), 1965 (1982)
53. Schnabl, H. et al.: Planta *149*, 280 (1980)
54. Scheurick, H. et al.: Biochim. Biophys. Acta *598* (3) 645 (1980)
55. Zenk, M. H.: Plant Cell Culture Symp., London, Oyez 1982
56. Gestrelius, S.: Enzyme Eng. *6* 345 (1982)
57. Vieth, W. R.: Ann. N.Y. Acad. Sci. *326*, 1 (1979)
58. Brillouet, J. M. et al.: Biotech. Bioeng. *18*, 1821 (1976)
59. Goldstein, W. E. et al.: Plant Tissue Culture as a source of biochemicals (Staba, E. J., ed.) p 191, Baton Rouge: C.R.C. Press 1980
60. Veliky, I. A., et al.: Biotechnol Lett. *3* (10) 551 (1981)
61. Galum, E. et al.: Planta Med. *49*, (1), 9 1983
62. Schuler, M. L.: Annals N.Y. Acad. Sci., *369*, 65 (1981)
63. Prenosil, J. E. et al.: Enzyme Mcrob. Technol. *5*, (5), 323, 1983
64. Brodelius, P. et al.: Eur. J. Appl. Microbiol. Biotechnol. *17*, (5) 275 (1983)
65. Schnabl, H. et al.: Planta *158*, (5), 392 (1983)
66. Schnabl, H. et al.: Physiol. Plant. *59* (1), 46 (1983)
67. Morris, P. et al.: Plant Cell Tissue Organ. Cult. *2*, 207 (1983)

Production of Useful Plant Metabolites

Masanaru Misawa
Allelix Inc., 6850 Goreway Drive, Mississauga, Ontario, L4V 1P1, Canada

1 Introduction

It is self-evident that the ultimate objective of research on the production of natural products by plant tissue and cell cultures is the industrial manufacture of useful plant metabolites on a large scale in a similar way to microbial fermentations. Unfortunately, this goal has not yet been realized except for the production of the shikonin derivatives. Recently, however, substantial progress has been made and some of the results are economically very promising.

In this review progress towards application is discussed together with problems which must be overcome from an industrial point of view.

2 Background

As described in earlier reviews [1-5], studies on the production of plant metabolites by callus and cell suspension cultures have been carried out on an increasing scale since the end of the 1950's. The large scale cultivation of tobacco and a variety of vegetable cells was examined in the late 1950's and early 1960's by Tulecke and Nickell at Pfizer Inc. [6], Mandels et al. at the Natick Laboratories in the U.S. Army [7], Street et al. at the University of Leicester [8] and Martin et al. at the National Research Council of Canada [9]. Their results stimulated more recent studies on the industrial application of plant tissue and cell cultures in many countries. Since Japan has a highly developed fermentation technology, many industrial companies, in collaboration with some university groups [10], have tried to apply this technique for the commercial production of useful compounds. However, in spite of many efforts most of these companies discontinued study in this field after the mid 1970's petroleum crisis because they thought the technique was not yet ready for commercial production. On the other hand, groups in West Germany outlined several very interesting approaches to industrial application [11] in a meeting held in 1976 at Munich, and their excellent results have encouraged researchers in other countries. In 1982, the 5th International Congress of Plant Tissue and Cell Culture was held in Japan and more than 700 participants from 46 countries attended the Congress, which was the largest one ever held. About 70 out of 372 papers presented there related to production of secondary metabolites in cultured plant cells and several papers to be commercially promising. Thus, research in this field was recently reactivated, particularly in Japan [12]. It should be noted that the academic studies carried out in universities and governmental institutes have also contributed to the progress in this field.

Higher plants contain a variety of substances which have been useful as medicines, food additives, perfumes etc. However, decrease of plant resources, increase in labour cost and other problems in obtaining these substances from natural plants have all strengthened the case for using plant cell culture. Because plant cell culture is not affected by changes in such environmental conditions such as climate, production is available in any place or season.

In spite of remarkable advances in plant tissue culture technology, the production cost of metabolites is still high, as estimated by Zenk [1] and Goldstein [13]. The former calculated that the cost would be U.S. $500 per kg of isolated product when it was accumulated in $1\ g\ l^{-1}$ within a period of 15 days in a $100\ m^3$ batch culture. It is true that the cost is decreasing since his estimation in 1974, and the producing ability of $1\ g\ l^{-1}$ for 15 days is reasonable in the case of some compounds such as rosmarinic acid, but it is still too expensive for the production of food stuffs, food additives or pharmaceuticals which can be more easily produced by chemical synthesis or by fermentation. Therefore, it is very important to choose what kind of products we should make through tissue culture technology.

In order to diminish the production cost, increase of production efficiency per cell is an essential factor. This means that higher amounts of product must be produced as quickly as possible. Many researchers have investigated this problem, and have tried various approaches using a variety of techniques and plant species, and some of them were very effective as I describe below.

3 Approaches to Increase Productivity

3.1 Optimization of Environmental Conditions

A variety of chemical and physical factors affecting cultivation have been tested extensively in many kinds of plant cells. These factors include media components phytohormones, pH, temperature, aeration, agitation, light, etc. There are so many reports and patents concerning this approach that it is impossible to discuss them in detail in this review, so several typical examples will be described instead.

Zenk et al. [14] tested various well-known basal media for the production of serpentine, an indole alkaloid, as summarized in Table 1. The results indicate that the amount of serpentine depends on the composition of the medium used. Among them, Murashige and Skoog's (MS) formulation was recognized to be the most suitable for the production of this particular alkaloid by *Catharanthus roseus* suspensions. This is of course not always true in other cultures.

Table 1. Effect of different media on growth and alkaloid (serpentine) formation in *Catharanthus roseus* cell suspension culture (Zenk et al. [14])

Basal medium[a]	Cell yield g dwt per l	Serpentine $mg\ l^{-1}$	Serpentine contents % dwt
Blaydes	7.6	4.4	0.06
Gamborg-B5; + 2,4-D: 1 $mg\ l^{-1}$	4.6	0.5	0.01
Gamborg + 2,4-D: 2 $mg\ l^{-1}$	5.2	0	0
Gamborg + NAA: 1.86 $mg\ l^{-1}$	7.6	1.2	0.02
Gamborg 4X	5.1	0	0
Heller + IAA: 0.175; BA: 1.13 $mg\ l^{-1}$	5.4	6.6	0.12
Linsmaier and Skoog	9.3	0	0
Murashige and Skoog	8.9	10.4	0.12
Nitsch and Nitsch	2.3	2.0	0.09
Veliky and Martin	5.0	0	0
White	0.8	0	0

[a] IAA = Indole-3-acetic acid;
NAA = 1-Naphthalene acetic acid;
2,4-D = 2,4-Dichlorophenoxy acetic acid;
kin = kinetin;
BA = Benzyladenine

Sucrose is the most popular carbon source for plant tissue culture and the level of the sugar also affects the productivity. For example, higher levels of sucrose stimulated the yield of rosmarinic acid by *Coleus blumei* [15] and of anti-plant virus substances by *Phytolacca americana* [16] and *Agrostemma githago* [17] cell suspension cultures (Figs. 1 and 2).

The type and concentration of inorganic nitrogen sources in the medium are also important factors for growth and productivity. As the author and his colleagues have shown [18], a higher concentration of nitrogen sources than that of MS medium stimulated the level of L-glutamine accumulated in *Symphytum officinale* cells, up to 20% of dry cell weight. The growth of tobacco cells was found to be much higher in a medium containing a higher level of nitrogen [19]. On the other hand, tobacco cells harvested from a nitrogen-rich medium were less suitable as a raw material for cigars and cigarettes because of the high (particularly protein) nitrogen concentration in the cells. To avoid this, the Japan Tobacco and Salt Public Corporation (now, the

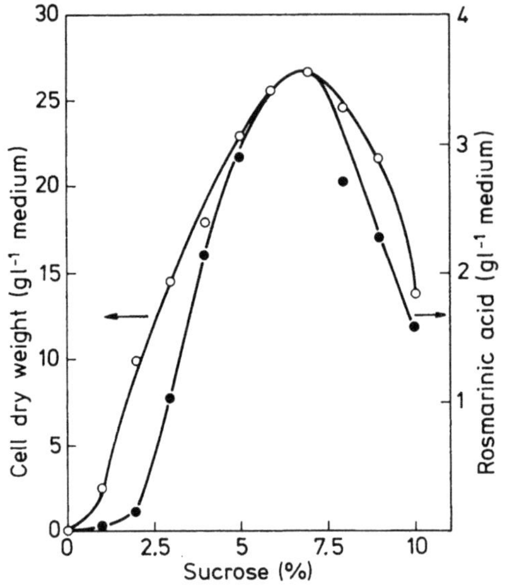

Fig. 1. Effect of sucrose level on rosmarinic acid production by *Coleus blumei* cell culture (Zenk et al. [15])

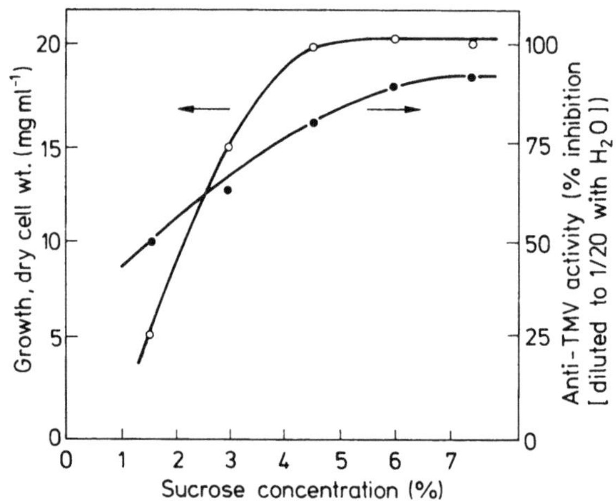

Fig. 2. Effect of sucrose level on anti-plant virus substances by *Agrostemma githago* cell culture (Takayama and Misawa [17])

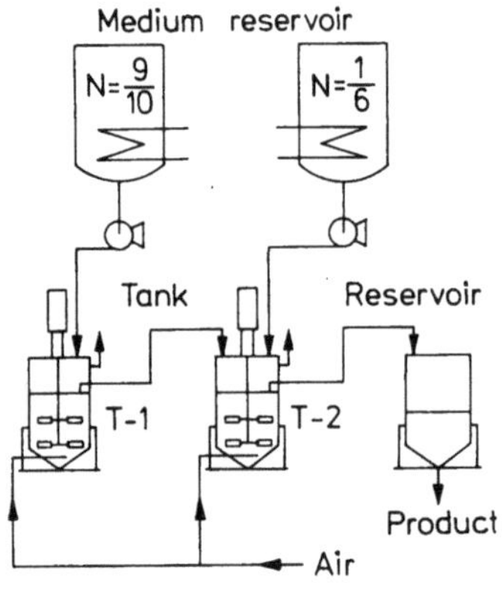

Fig. 3. Two stage and two stream culture for tobacco cell culture (Noguchi et al. [20]).
T-1: culture tank I
T-2: culture tank II

Japan Tobacco Industry Co.) has developed a "two stage and two stream cultivation method" as shown in Fig. 3 [20]. In this system, fresh MS medium containing 9/10 of the normal nitrogen source and 3 times stronger phosphate was continuously supplied in Tank I and the inflow was balanced by the outflow of the corresponding volume of the culture which was transferred to Tank II. The latter was supplied with another MS medium with 1/6 of the normal nitrogen source and 3 times stronger phosphate. Using this system, they obtained cells continuously at a rate of 6.9 g dry wt. per l. Using a 20 m^3 vessel they could cultivate cells continuously for 66 days and the cell level was 16.5 g l^{-1} with a production rate of 5.85 g l^{-1} day [21]. In the case of a batch culture, a very rapid doubling time (12 h) was achieved in the nitrogen-rich medium, the fastest growth rate recorded for plant cell suspension cultures.

When a product contains nitrogen in its molecule, the concentration of inorganic nitrogen sources in the medium sometimes affects its level in the cells. The author found that the amounts of proteinase inhibitors accumulated in *Scopolia japonica* cells [22] and of anti-plant virus substances in *P. americana* [16] cells were markedly affected by the levels of nitrogen compounds in the medium. Both of these compounds are proteinaceous.

Among various inorganic salts in the basal medium, a higher level of phosphate often stimulates cell growth, as reported by Kato et al. [23] using tobacco cells. Although phosphate showed a stimulative effect on growth, it decreased the level of alkaloids in tobacco, *Catharanthus roseus* and *Peganum harmala* cultured cells [24]. According to the results, cells of these plants produced higher amounts of ajmalicine and phenolics (*C. roseus*), caffeoyl putrescine (*N. tabacum*) and harman alkaloids (*P. harmala*) in induction media whose common characteristics were the lack of phytohormones and inorganic phosphate (Pi). However productivity decreased markedly when the cells

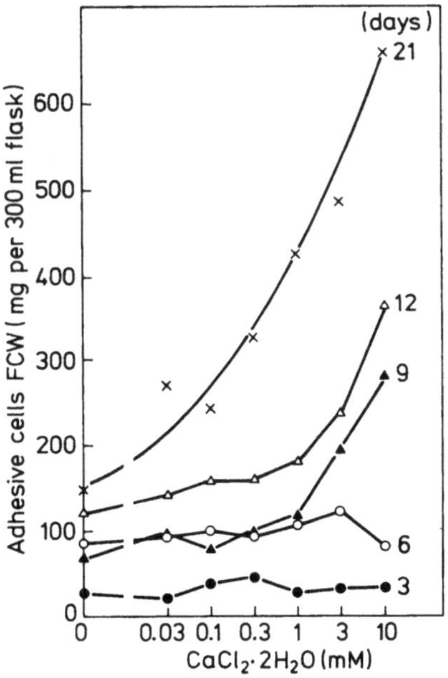

Fig. 4. Effect of $CaCl_2 \cdot 2\,H_2O$ on cell adhesion of *Agrostemma githago* cell suspension culture [17]. Adhesion of cells on the surface of a 300 ml Erlenmeyer flask

were transferred to induction media containing 1.2 milliatom Pi. At the same time, it was found that activity of enzymes diverting the corresponding primary precursors into secondary pathways, such as phenylalanine ammonialyse, was reduced in the Pi containing medium. Berlin and co-workers [24] suggested that either Pi or phosphate metabolites might be involved in the regulation of secondary metabolism. Similar phenomenona are known from antibiotic production by microorganisms.

Ca^{2+} and other major or minor elements in basal medium may also have an affect on growth and productivity. As a result of studies on cultural condition to produce digitoxin using *Digitalis purpurea* shoot forming cultures, Hagimori et al. [25] observed that reduction of $CaCl_2$ to one-third, addition of 3 mM $MgCl_2$ and 3-fold increase of KI, Na_2MoO_4 and $CuSO_4$ in MS medium improved digitoxin formation. The authors and his colleagues [26] found that the number of *A. githago* cells adhering to the wall of the fermentor was markedly reduced, and the cells were easily removed from the inside wall of a fermentor to the medium, by reducing the level of $CaCl_2 \cdot 2H_2O$ in MS medium (Fig. 4). Foaming was also suppressed in the low Ca^{2+} medium. This indicates that the adhesion of cells seemed to be caused by gel formation from pectin and Ca^{2+}. These results are particularly interesting for large scale cultivation.

pH in the medium also affects growth and production [3]. Since it is difficult to control the pH of the medium in a flask scale experiment, its effect has been mainly examined using jar fermentors and big tanks. In tripdiolide production by *Tripterygium wilfordii*, the level of production of this antineoplastic agent was higher at pH 6.0 than pH 7.0 [27].

Among a number of components in the medium, phytohormones have the most remarkable effects on growth and productivity. In general, increase in auxin levels, such as 2,4-D, in the medium stimulates dedifferentiation of the cells and consequently diminishes the level of secondary metabolites. However, it is not always true since production of L-DOPA by *Mucuna pruriens* [28], of ubiquinone-10 by *N. tabacum* [29] and of diosgenin by *Dioscorea deltoidea* [30] was stimulated by high levels of 2,4-D. Ambid et al. [31] were able to increase the amounts of ajmalicine and serpentine using both growth medium and production medium. The former contained 2,4-D while the latter did not. Okazaki et al. [32] investigated the effect of phytohormones in detail on the production of scopoletin and scopolin and Ikuta et al. [33] examined the effect on alkaloid production of different 2,4-D levels in the medium. Using callus tissue

Table 2. Stimulation of tripdiolide production by addition of N-phenyl-N^1-(2-chloropyridyl) urea in *Tripterygium wilfordii* cultures [27]

	Level $mg\,l^{-1}$	Growth $mg\,l^{-1}$ dry wt.	Tripdiolide $\mu g\,l^{-1}$
4-PU-30	0.01	12.5	72.5
	0.1	11.7	153.5
	1.0	6.8	45.0
Kinetin	1.0	10.8	91.1

Table 3. A modified medium for production of anti-plant virus substrates by *Agrostemma githago* cell cultures

Components	$mg\ l^{-1}$	Components	$mg\ l^{-1}$
NH_4NO_3	1650 (20.6 mM)	KI	0.83
KNO_3	1900 (18.8 mM)	$Na_2MoO_4 \cdot 2\ H_2O$	0.25
$CaCl_2 \cdot 2\ H_2O$	22 (0.15 mM)	$CuSO_4 \cdot 5\ H_2O$	0.025
$MgSO_4 \cdot 7\ H_2O$	370 (1.5 mM)	$CoCl_2 \cdot 6\ H_2O$	0.025
KH_2PO_4	340 (2.5 mM)	Thiamine-HCl	0.4
$Na_2EDTA \cdot 2\ H_2O$	37.3	2,4-D	0.1
$FeSO_4 \cdot 7\ H_2O$	27.8	Sucrose	30,000
H_4BO_3	6.2	Water	1000 ml
$MnSO_4 \cdot 4\ H_2O$	22.3	pH	6.3
$ZnSO_4 \cdot 7\ H_2O$	10.6		

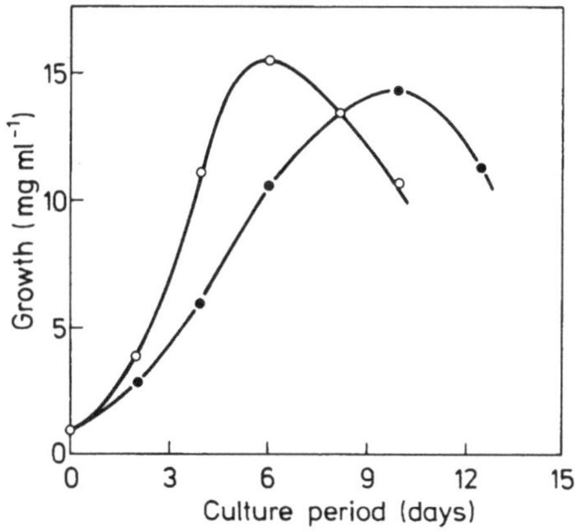

Fig. 5. Time courses of *Agrostemma githago* cultures in original and modified media.
● original MS medium
○ modified MS medium (see Table 3)
3% sucrose, 0.1 $mg\ l^{-1}$ Kinetin and 0.05 $mg\ l^{-1}$ 2,4-D were used

of *Thalictrum minus*, they isolated a large amount of berberine, six protoberberine type alkaloids and an aporphine type alkaloid.

4-Chloro-2-diphenylurea (4-PU) is a new type of compound having cytokinin activity [34]. As seen in Table 2, a suitable amount of 4-PU stimulates the synthesis of tripdiolide by *T. wilfordii* cell cultures [27].

As shown by several reports, it is possible to increase production ability by optimizing cultural conditions. Using the modified medium shown in Table 3, cell growth became faster and the level of anti-plant virus substances increased in *A. githago* [26] (Fig. 5). Fujita et al. [35] modified White's medium to a production medium, M-9, in order to produce higher amounts of shikonin derivatives by *Lithospermum erythrorhizon* suspension cultures. The cells produced 120 $mg\ l^{-1}$ shikonin derivatives in White's medium but 1400 $mg\ l^{-1}$ in M-9 medium as seen in Tables 4 and 5.

They established a two-stage culture method using this modified system in which the cells were firstly placed in growth medium, and subsequently transferred into production medium. This approach was used with a 200 l air-lift reactor fitted with an agitator for the growing stage, and then a similar 750 l reactor for the production stage. As a result, 1.5 g of shikonin derivatives per litre were obtained and

Table 4. Compositions of White's medium and M-9 medium

Component	White's medium (1954)	M-9 medium
	mg l^{-1}	mg l^{-1}
$Ca(NO_3)_2 \cdot 4\ H_2O$	300	694
KNO_3	80	80
$NaH_2PO_4 \cdot 2\ H_2O$	21	19
KCl	65	65
$MgSO_4 \cdot 7\ H_2O$	750	750
Na_2SO_4	200	1,480
$MnSO_4 \cdot 4\ H_2O$	5	—
$ZnSO_4 \cdot 7\ H_2O$	3	3
$Fe_2(SO_4)_3$	2.5	—
NaFe-EDTA	—	1.8
H_3BO_3	1.5	4.5
KI	0.75	—
$CuSO_4 \cdot 5\ H_2O$	0.01	0.3
MoO_3	0.001	—
Sucrose	20,000	30,000
Glycine	3	—
Nicotinic acid	0.5	—
Thiamine · HCl	0.1	—
Pyridoxine · HCl	0.1	—

Table 5. Cell growth and production of shikonin derivatives by *Lithospermum erythrorhizon* cell cultures in White's and M-9 media [35]

Medium	Cell yield (g DW per l)	Shikonin derivatives formed (mg l^{-1})	Amount of shikonin derivatives (%)
M-9[a]	11.3	1400	12.4
White[b]	5.7	120	2.1

[a] Supplemented with 10^{-5} M 3-indoleacetic acid;
[b] Supplemented with 10^{-6} M 3-indoleacetic acid and 10^{-5} M kinetin

the level corresponded to 14% dry weight cells. A major component of the derivatives was acetylshikonin. Furthermore the ratio of several other components in the product was recognized to be similar to those in the roots of the plant [36]. The product is being produced as a pigment for a lip-stick. A patent [37] was also filed describing addition of cellulose fibre, liquid paraffin or fatty acids to the liquid medium to stimulate the amount of shikonin derivatives.

As seen in the reviews by Martin [9], Fowler [3] and others, physical conditions for cultivation such as aeration, agitation and so on are also very important factors, particularly in large scale process. In anthraquinone production by *Morinda citrifolia*, Wagner [38] suggested an air-lift type reactor was most suitable for plant cell culture. Biotransformation of β-methyldigitoxin to β-methyldigoxin by immobilized cells of *Digitalis lanata* is being studied using both Erlenmeyer flasks and bubble columns by Alfermann et al.[40]. No difference in productivity was recognized by them. Using 7 l air-lift type reactors, two-stage semi-continuous culture of *C. roseus* for production of serpentine was compared with a batch culture by Smart et al.[41] and the latter, with a low level of sugar, was recognized to be advantageous. However, continuous culture is not always the best for cell growth, for example, Tal et al.[42] reported that 1.8% d.w. of diosgenin was obtained in 18 days in batch culture but the amount was much lower in continuous culture.

Light irradiation is sometimes effective for secondary metabolite production[43]. Light-grown, shoot-forming cultures of *Digitalis purpurea* were found to accumulate considerable amounts of digitoxin, approximately 20 to 40 microorganisms per g dry weight being typical. On the other hand, white, shoot-forming cultures without chloroplasts produced only one-third the amount of digitoxin as seen in Fig. 6.

Recently DiCosmo et al.[44] reviewed possible correlations between stress and secondary metabolism in cultured plant cells. In particular effects of fungal induced stress on phytochemical synthesis was described. Compounds which induce stress-metabolites, named elicitors, such as glucan in cell walls of *Phytophthora megasperma*, were found to stimulate to some extent the synthesis of certain metabolites and related enzymes. The use of appropriate elicitors to increase the productivity of secondary metabolites might be a useful approach.

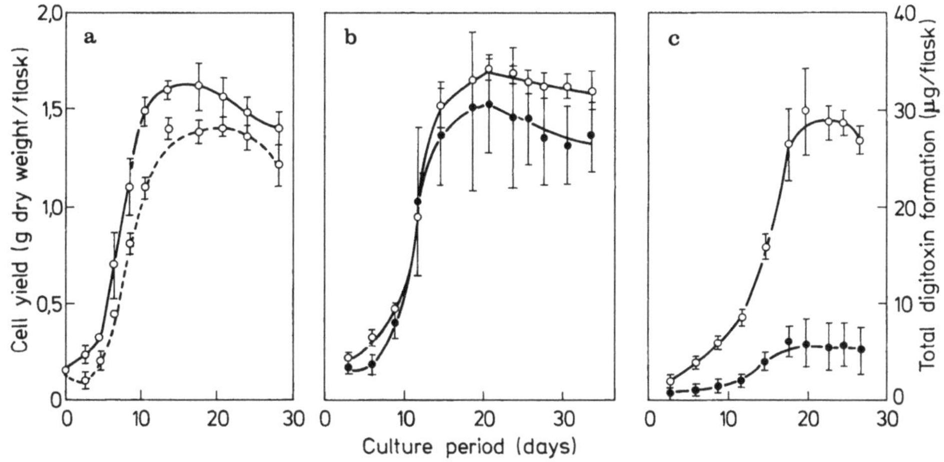

Fig. 6a–c. Growth and digitoxin formation of *Digitalis purpurea* L. cell culture (Hagimori et al.[43]).
a Growth curve of undifferentiated cells, O——O green cells; O---O white cells.
b Growth and digitoxin formation of green, shoot-forming cultures, O——O growth; ●——● digitoxin content.
c Growth and digitoxin formation of white, shoot-forming cultures, O——O growth; ●----● digitoxin content

3.2 Addition of Precursors and Biotransformation

Addition to the culture media of appropriate precursors or related compounds sometimes stimulates production ability and this approach is advantageous if the precursors are inexpensive. Since Chan and Staba[45] initially examined the production of alkaloids with this approach in the 1960's, many similar experiments have been carried out. For example, amino acids have often been added to cell suspension culture media for production of tropane alkaloids, indole alkaloids, and ephedrin. Some stimulative effects have been observed. It is true that some amino acids are precursors of various alkaloids, but generally the biosynthetic steps from amino acids to alkaloids are so complicated that the author doubts whether the amino acids added were incorporated into the alkaloids in cell culture.

Perhaps they affected not only alkaloid biosynthesis directly as precursors, but also indirectly through other metabolic pathways in the cells.

Tabata et al.[46] reported that addition of 500 mM tropic acid to the medium of *Scopolia japonica* increased the amount of alkaloids by up to 14 times. The level of an antineoplastic agent, tripdiolide, was increased by addition of 100 μg l^{-1} farnesol[27,141]. (Table 6) Farnesol is a dephosphorylated compound of farnesyl pyrophosphate which is an intermediate in the biosynthesis of terpenoids.

Table 6. Effect of addition of farnesol on production of tripdiolide by *Tripterygium wilfordii* cell cultures[27]

Farnesol add μg ml^{-1}	Growth mg ml^{-1}	Tripdiolide μg per g cell dry wt.
0	14.3	6.3
30	12.7	13.0
100	10.1	20.4

Cultivated for 3 weeks

Biotransformation has been extensively applied in the fermentation industry, for example in the production of L-aspartic acid from fumaric acid, L-malic acid from the same substrate and interconversion of a variety of steroids. Using the same approach plant cell cultures have been used in studies on the formation of steroids and alkaloids on a laboratory scale, but industrial application has not yet been achieved[47].

Biotransformation of β-methyldigitoxin to β-methyldigoxin using *Digitalis lanata* cells has been a possibility since 1974 following the work of Reinhard and Alfermann[48]. Digoxin has a big market as a cardiac glycoside. To reduce the production cost, which is very important in industrial application, Reinhard et al. examined the use of immobilized cells[49] and semi-continuous culture[50]. The result of a typical biotransformation for 17 d using a 20 l reactor is shown in Table 7. In this reaction, the cell growth was 12–14 g l^{-1} and the maximum level of product was 700 mg l^{-1}. The scale-up experiments using a 200 l reactor was carried out by using, in sequence, 20 ml flasks — (7 d), a 35 l air-lift reactor — (10 d) and a 200 l air-lift reactor — (18 d) cultivations. As a result, about 600–700 mg of β-methyldigoxin per litre was obtained using a 200 l reactor. The biotransformation is now being studied by Boehringer Mannheim Co., using 4000 l tanks.

Table 7. Biotransformation of β-methyldigitoxin to β-methyldigoxin by *Digitalis lanata* cells in a 20 l reactor

β-Methyldigitoxin added	17.24 g	(100 %)
Unconverted β-methyldigitoxin	2.04 g	(11.8%)
β-Methyldigoxin formed	14.36 g	(81.7%)
By-product	0.28 g	(1.4%)
Yield	94.90%	

Kurz et al. [51] found that 3′,4′-dehydrovinblastine was converted to leurosine and catharanthine, which also accumulate in *C. roseus* cells, with conversion rates of 31% and 9%, respectively by a selected strain of *C. roseus*. This group also reported the biotransformation of thebaine to neopine by *Papaver somniferum* cv. *Marianne* cell suspension cultures [52].

A culture of *Daucus carota* was shown to specifically hydroxylate digitoxigenin at the C-5 position and also to convert gitoxigenin and oleandrigenin to their corresponding 5β-hydroxy derivatives in a high yield by Jones and Veliky [53]. Lui and Staba [54] with leaf liquid cultures and root cultures of *Digitalis lanata* found that the former grew very rapidly as compared to the root cultures and the unorganized cell suspension cultures. The concentration of digoxin increased in both leaf and root cultures by addition to the medium of either sodium glycocholate, cholesteryl acetate, or progesterone.

Hiraoka [55] reported the conversion reaction of 2-(p-methoxy phenyl) ethylamine to 2-(p-hydroxyphenyl) ethylamine. On the other hand, this substrate was transformed to 2-(p-methoxy phenyl) ethyl-β-D-glucopyranoside by *Strobilanthes dyerianus*.

To produce a prodrug. Umetani et al. [56] have studied the glucosylation of umbelliferon and salicylic acid using plant cells. They found that *Mollotus japonicus* had the ability to convert salicylic acid to its o-glucoside with a yield of 90–95%. The maximum level of the product obtained was 0.9 g l^{-1}. It is of interest that the glucoside showed as potent an analgesic activity as salicylic acid, while its effect was more rapid and more long-lived than that of salicylic acid in mice (Fig. 7).

The conversion of geraniol and nerol to neral and geranial by *Vitis vinifera* cell suspension cultures is also interesting. Using plant cell culture techniques, radio-

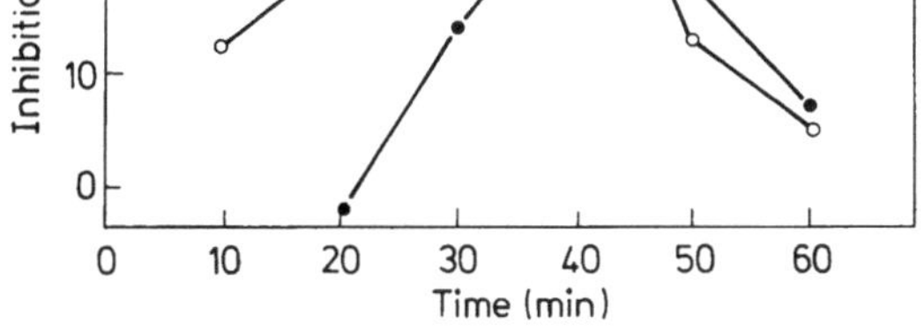

Fig. 7. Analgesic activities of salicylic acid and its glucoside (Umetani et al. [56])

activity labelled compounds can be formed from appropriate substrates as reported by Mangold et al. [58]. This is a useful application of the technology because of the high value of the product.

3.3 Selection of High-producing Strains

The physiological characteristics of individual cells are not always uniform. For example, pigment producing cell aggregates typically consist of producing cells and non-producing cells. In 1976, Zenk [59], obtained cell lines of *Catharanthus roseus* which could produce high levels of ajmalicine and serpentine as determined by radioimmunoassay. Following their excellent results, a number of researchers have used cell cloning methods as this is the most promising way of increasing the levels of metabolites present. Some typical examples are shown in Table 8. Most of them are related to production of pigments, such as anthocyanins, as selection is easy because of their colour. Yamakawa et al. [60] isolated a strain of *Vitis* hybrid using an agar plating feeder method [61] which could produce 3.4% d.w. of anthocyanin. They also tried to isolate high-producing cells through protoplast culture, but almost all of the colonies as obtained were white cells. A strain of *Euphorbia millii* was also recognized to accumulate about 7 times higher amounts of anthocyanin than that of the parent strain after 24 selections [62]. Statistical and cell-pedigree analysis proved that this strain produces the red pigment stably.

According to Constabel's investigation [63] with *C. roseus*, anther, leaf and meristem explants revealed that synthesis and accumulation of alkaloids in cell cultures differ significantly provided the comparison is based on individual alkaloid types. Variation of alkaloid profiles in callus tissues developed from the anther walls and fillaments of *C. roseus* ranged from cell lines with no detectable alkaloids to those with 12 alkaloids representing Corynanthe-, Strychnos-, Aspidosperma-, and Ibogatype alkaloids [64].

Berberine is useful as an intestinal antiseptic in the Orient. Since it takes 5–6 years to produce *Coptis* roots as the raw material, Yamada et al. [65,66] selected a high berberine producing cell line of *Coptis japonica* and cultivated the strain in a 14 l bioreactor. After induction of the callus, they repeated cell cloning using cell aggregates and isolated a strain which grew faster and produced a higher amount of the alkaloid.

The cells proliferated about 5 to 6 times in 3 weeks and the highest amount of berberine activated was 13% in dry cell material corresponding to 1.2 g of berberine per l of medium. The selected strain was quite stable, producing a high level of the

Table 8. Typical examples of cell cloning application in Japan

Products	Plants	Factor	Researchers
Anthocyanins	*Vitis* hybrid	2,3-4	Yamakawa et al.
Anthocyanins	*Euphorbia milli*	7	Yamamoto et al.
Berberin	*Coptis japonica*	2	Satoh et al.
Biotin	*Lavendula vera*	9	Watanabe et al.
Ubiquinone-10	*Nicotiana tabacum*	15	Matsumoto et al.

alkaloid even after 27 transfers. Since berberine is an interesting compound as a drug, its production has been examined by many researchers: Furuya et al. [67] identified it in a callus of *C. japonica* in 1972; Fukui et al. [68] also cloned a high producing cell line of *C. japonica*; Khanna et al. [69] reported an increase in the level of the alkaloid by addition of tyrosine in the medium using *Argemone mexicana* cell cultures, and Ikuta et al. [70] isolated a large amount of berberine, six protoberberine-type alkaloids and one aporphine-type alkaloid from callus tissue derived from the stem of *Thalictrum minus*. However, the price of berberine is not high enough for it to be produced by cell cultures.

Production of shikonin derivatives by *Lithospermum erythrorhizone* has been studied by Tabata's group since 1974 [71, 72]. The naphthoquinone pigments in this plant are traditionally used for the treatment of burns, skin diseases and hermorrhoids in Japan and in China (Fig. 8). Fujita et al. [35, 36] worked on large scale production of shikonin derivatives using Tabata's strains described above. During the course of their experiments, the cultured cells were treated with 0.5% Macerozyme, 1.0% Driselase and 2.5% cellulase to prepare protoplasts, and cell colonies were obtained from protoplasts on agar media. As a result of cultivation in production media, 15 strains out of 48 produced higher amounts of shikonin derivatives than those of the parent strain. The maximum productivity was twice as high as the parent, which was also higher than 4 g per g seed in 23 days.

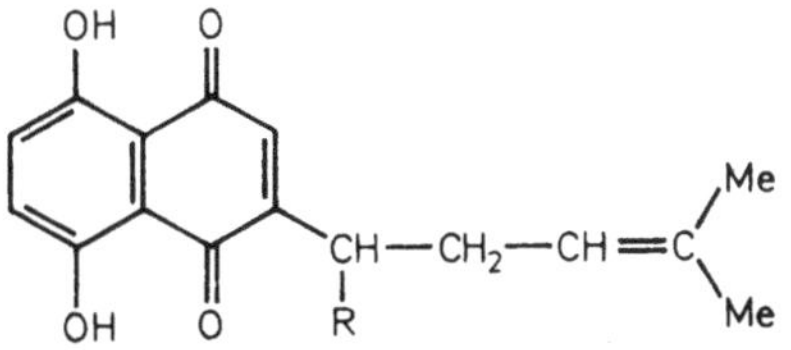

R:	
OH	shikonin
H	deoxyshikonin
OCOMe	acetylshikonin
$OCOCHMe_2$	isobutylshikonin
$OCOCH{=}CMe_2$	β·β-dimethylacrylshikonin
$OCOCH_2CHMe_2$	isovalerylshikonin
$OCOCH_2CH(OH)Me_2$	β-hydroxyisovalerylshikonin

Fig. 8. Chemical formula of shikonin and its derivatives

Cultured green *Lavendula vera* cells grown in the light were found to accumulate a high level of free biotin by Watanabe et al. [73, 74]. They used pimelic acid, a precursor of biotin, which has a toxic effect on cultured cells of *L. vera*, as a selecting agent to screen for high-producing cell lines. A strain obtained produced approximately 0.9 $\mu g\ g^{-1}$ fresh weight biotin which was 15 times higher than that of the parent strain. The level was also 10 times the amount found in the leaves.

Selection of a high ubiquinone-10 producing *Nicotiana tabacum* strain has produced dramatic results [75, 76, 77]. A group in the Japan Tobacco and Salt Public Corporation isolated a number of strains producing high levels of ubiquinone-10 from tobacco cell cultures and analyzed the contents of the cells by high performance liquid chromatography and by *Craven* assay. Several strains producing large amounts of ubiquinone-10 in suspension culture were selected and were subjected to further cell cloning. After the 13th recloning, a strain was selected from approximately 4000 cell clones tested giving a ubiquinone-10 level of 5.2 mg per g dry weight of cells. When *N. tabacum* BY-2, a parent of strain used for the cloning, was isolated as a producer in 1976, the titer for ubiquinone-10 was only 360 μg g^{-1} d.wt., therefore, the level was increased by more than 14 times by selection. It may be noted that 5.2 mg g^{-1} corresponds to 180 times the amount of that present in the mother plant. Most ubiquinone-10 has been commercially produced by a bioprocess as a drug for heart diseases in Japan and it is unlikely that plant cell cultures will replace microbial methods at present because of the cost.

Recently, Zieg et al. [79] reported the selection of high pyrethrin producing tissue cultures derived from *Chrysanthum cinerariaefolium* and indicated that analytical screening of the tissue lines enabled the selection of a few "high yielding" strains which were derived from high yielding plant selections. A rapid assay method is crucial in the selection of a high yielding cell line. Deus and Zenk [80] reported that a procedure using fluorescence assay to select ajmalicine and other major heteroyohimbine alkaloids producing cells of *C. roseus* was of clear advantage even over a radioimmunoassay procedure since almost unlimited numbers of colonies could be screened in this way.

Cell cloning is undoubtedly a very useful technique to increase the level of secondary metabolites and it should be applied as widely as possible. However, it is not obvious why cultures contain both high- and low-yielding cells. Only a few papers concerned with possible mechanisms have appeared. Böhm [81] indicates that the lack of specific enzymes(s) represents the most important reason for the inability of plant cell cultures to produce secondary metabolites. Berlin et al. [82] compared the highest cinnamoyl putrescine producing, p-fluorophenylalanine resistant strain TX4 of *N. tabacum* L. CV Xanthi with a low producing strain for five enzymes of the biosynthetic pathway. As a result, activities of these enzymes, phenylalanine ammonia-lyase, trans-cinnamate-4-hydroxylase, 4-coumarate: COA ligase, ornithine decarboxylase and arginine decarboxylase were found to be 3 to 10 times higher in TX4 cells. This sort of study will be very useful to isolate mutants able to overproduce useful plant metabolites.

3.4 Induction of Mutants

In biotechnology, induction of genetic mutant strains of microorganisms is ubiquitous and auxotrophic and regulatory mutants are used extensively to produce amino acids, nucleotides, antibiotics etc. However, vegetative plant cells are normally diploid and it is much more difficult to obtain genetic mutants. As Widholm indicated [83], most of the variants found so far in cultured cells cannot be classified as being due to mutation or nonmutation. Furthermore, biosynthetic pathways of many secondary metabolites and their regulation mechanisms in higher plants are obscure, therefore, it is also difficult to decide what kind of mutants should be induced in order to increase production ability.

However, Berlin [86] induced p-fluorophenylalanine (PFP) resistant cell lines of tobacco cell cultures and found that five lines of *Nicotiana tabacum* L Xanthi, one of *N. tabacum* L CV Maryland Mammoth and five of *N. glauca* out of 31 resistant cell lines accumulated higher levels of phenolics. He and his group [87] also reported that *N. tabacum* TX4, which is resistant to PFP, produced 6 to 10 times higher levels of cinnamoyl putrescine than that of the parent strain, TX1. Figure 9 shows the comparison of enzyme activities involved in the biosynthesis of cinnamoyl putrescines

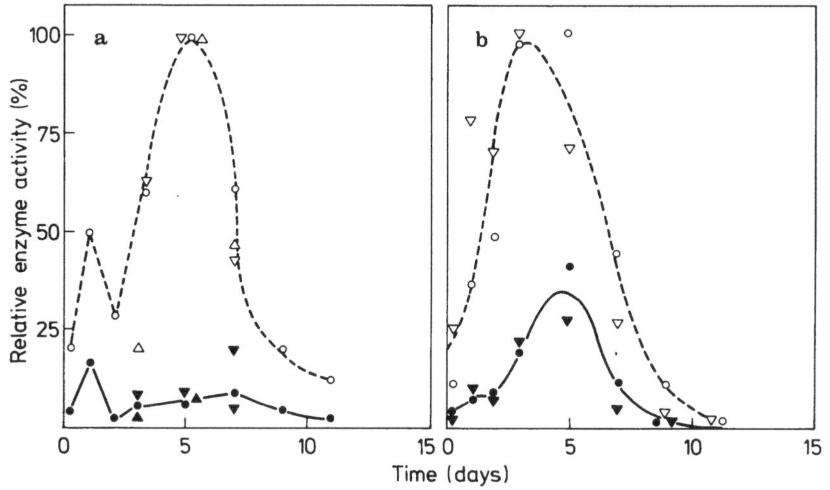

Fig. 9. Comparison of enzyme activities involved in the biosynthesis of cinnamoyl putrescines in TX1 (filled symbols) and TX4 (open symbols) of *N. tabacum* (Berlin et al. [86])

in TX1 and TX4. Enzyme activities in the PFP resistant strain were much higher than those of the parent. They [88] also found that a parent strain of *C. roseus* produced catharanthine only in the production medium but its tryptophan analog resistant mutant accumulated the same alkaloid even in the growth medium. Among several types of analogues tested, 4-fluorotryptophan was found to be the most eficient of them.

Increase of metabolite levels using regulatory mutants is theoretically possible and selection of suitable analogues for this purpose could be an important factor in order to produce a variety of products.

Several research groups [89-93] have used mutagens such as N-methyl-N^1-nitro-N-nitrosoguanidine, ethylmethane sulfonate, N-ethyl-N-nitrosourea or X-rays to induce high yielding strains.

In China [94], a fine callus strain of *Anisodus acutangulus* was derived through irradiation with 4000 R of X-ray and the level of scopolamine in the cells was 0.177 ± 0.001 mg g^{-1} d.wt., which was about 30% higher than that of the parent. The productivity was stable.

Deus [95] treated *C. roseus* cells with X-rays and obtained a cell line having a high producing activity of serpentine whose level was 2%.

A cell line of *Lavendula vera* producing a high level of free biotin was obtained by gamma ray irradiation for 1 h with ^{60}CO (dose 10 KR h^{-1}) by Watanabe et al [73]. The line was found to contain 7 times the amount, 0.425 μg g^{-1} fresh wt., of free biotin compared with the original unselected cells, 0.061 μg g^{-1} fresh wt., and 4.5 times that found in the leaves.

In spite of the difficulty of obtaining auxotropic mutants in plant cells, several groups have been actively working in this approach using haploid cells or protoplasts [96, 100] although their aim is not always plant metabolite production. This approach could be valuable.

Induction of a mutant having altered permeability could also be important, because plant cells generally accumulate their metabolites intracellularly. This is very disadvantageous in commercial production because the amount of compounds produced is limited to a low level. If the cells could excrete them in large amounts, the product cost would be reduced. According to Berlin's results [101] *Thuja occidenta* excreted monoterpenoids but the levels in the medium were only 5% of those in the mother plant. The same group reported that *Macleya microcarpa* cells excreted nearly all the alkaloids detectable in the culture flask [102]. Yamakawa et al. [103] cultivated *Tinospora rumphii* cells in 14 ml medium and found 0.57 mg (5.3% in dry weight cells) of isoquinoline alkaloids in the cells and 0.50 mg in the culture filtrate after 7 d cultivation. By replacement of the medium with fresh medium after 3 d of cultivation, 0.50 mg of the alkaloids in the cells and 1.02 mg in the filtrate were accumulated. It seems then that the products must be excreted to some extent. Unfortunately, the secretion mechanisms for secondary products in higher plant cells have not yet been elucidated and extensive fundamental studies are required before meaningful manipulation can take place.

3.5 Application of Immobilized Cells

Since the growth rate of most plant cells is generally lower than that of bacteria, culture must proceed for longer. Use of immobilized cells instead of free cells may therefore be advantageous as indicated by Brodelius and Mosbach [104]. Compared

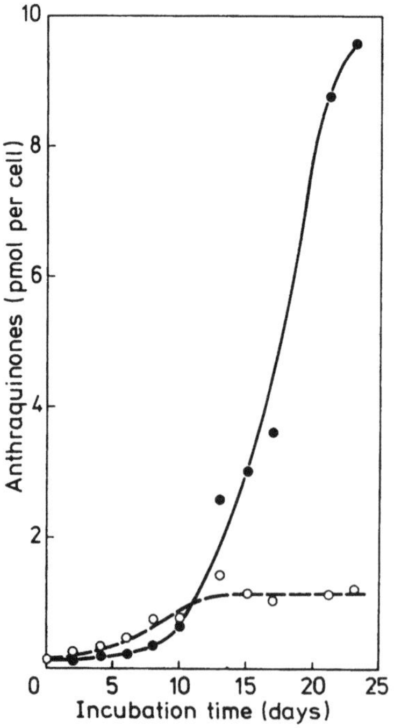

Fig. 10. Anthraquinones synthesis by freely suspended (○) and immobilized (●) cells of *C. roseus* (P. Brodelius et al. [105])

with microorganisms, only a few groups have worked on the application of this technique for plant cells. Brodelius et al. [105, 106] used *Morinda citrifolia* cells immobilized by entrapment in alginate to produce anthraquinone. As seen in Fig. 10, it was found that the immobilized cells had an increased synthesis of anthraquinones as compared to freely suspended cells under the same conditions. They also activated the synthesis of indole alkaloids (10% yield) from tryptamine and secologanine using immobilized *C. roseus*. Although the products are normally stored in the vacuoles of the cells, addition of a trace amount of chloroform in the aqueous phase altered cell permeability and consequently the products were excreted extracellularly. In a comparison with several types of polysaccharides as matrices, they recognized that alginate entrapped *C. roseus* had the highest productivity of ajmalicine [104] and that respiration rates and rates of cell growth of *Glycine max* cells entrapped by agar or agarose were approximately the same as those of freely suspended cells [107] (Fig. 11).

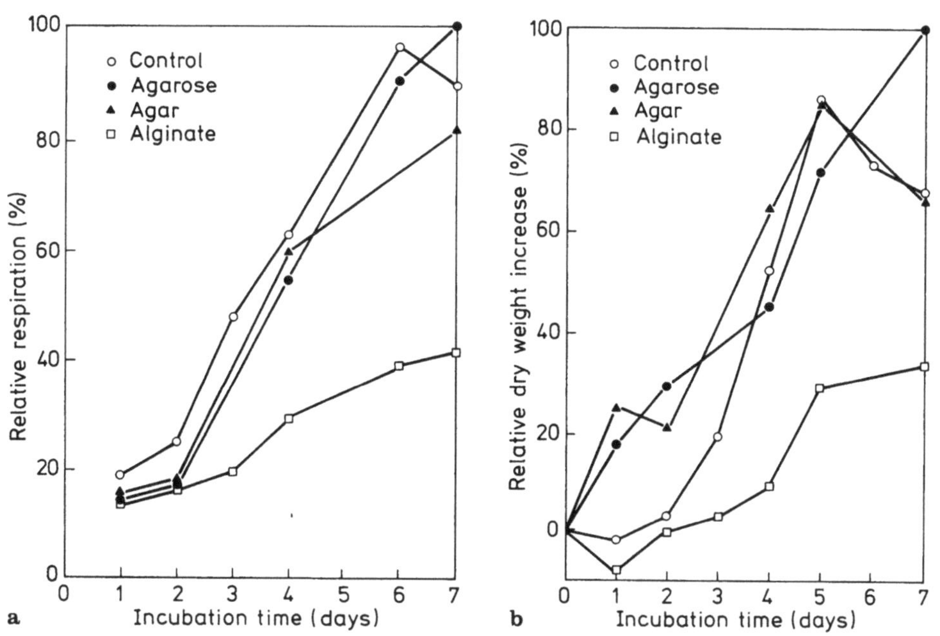

Fig. 11. Respiration (**a**) and cell growth (**b**) of *Glycine max* cells in free suspension and immobilized in various polysaccharides (Brodelius et al. [107])

Biotransformation from β-methyldigitoxin to β-methyldigoxin using alginate entrapped *D. lanata* has also been studied by Alfermann et al. [108]. Biotransformation occurred at a constant rate for at least 150 d after a lag phase of about 20 d. Furthermore, they tested cultivation of the entrapped cells in a bubble column similar to that used by Veliky et al. [109]. No difference in the productivity was found between bubble columns and Erlenmeyer flasks as seen in Fig. 12. Using immobilized carrot cells, Veliky and Jones [109] reported 5-β-hydroxydigitoxigenin formation from gitoxigenin in the yield of 80%. The activity continued for more than 30 days.

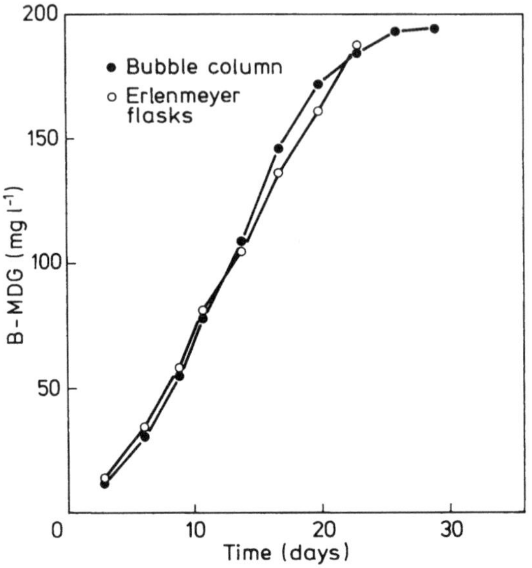

Fig. 12. Biotransformation of β-methyldigitoxin to β-methyldigoxin by immobilized cells of *Digitalis lanata* in erlenmeyer flasks and in a bubble column [108]. The cells are cultivated continuously. The medium is changed at the days indicated, and new substrate (40 mg l^{-1}) is added

Yoshikawa et al. [110] recently reported that Ca-alginate was superior to other entrapping materials for a variety of plant cells. The immobilized cells of *Papaver somniferum* survived for 6 months when the medium was replaced with fresh medium every 2 months, but immobilized cells of *Nicotiana tabacum*, *D. lanata*, and *Glycyrrhiza echinata* were excreted from gel matrices after the 10th day of cultivation. The entrapped cells of tobacco and Digitalis showed glucosylation and palmitylation activities of steroids as high as free cells, but did not excrete products into the medium. On the contrary, immobilized *P. somniferum* (opium poppy) cells converted codeinone to codeine (70% yield) and excreted most of the morphinan alkaloid. This system was also used successfully in a continuous reaction.

3.6 Morphological Differentiation and Productivity

It is, of course, desirable to use morphologically undifferentiated cells for the production of a large amount of useful metabolites in ways similar to microorganisms and there are many examples which show high productive ability of such cells compared with intact plants as already described in this review. On the other hand however, there are also large numbers of reports describing the loss of secondary metabolite productivity in undifferentiated cells. Therefore, studies have been carried out using differentiated cells which have roots, shoots or other organs. As discussed by Wiermann [111], it is generally accepted that in the intact plant a close correlation exists between the expression of secondary metabolism and morphological and cytological differentiation. However, in cultured cells, it is not clear whether this correlation still holds. The author has a callus of *Phytolacca americana* which was induced from stem tissue more than 10 years ago. It looks morphologically undifferentiated but still produces a high amount of betanine, an alkaloid. Wiermann concluded that it is not yet clear to what extent secondary metabolism depends on the development of specific structures, and it is unknown whether these two processes are genetically and/or physiologically linked.

Krueger et al. [112] have cultured tissues possessing leaf organs from *C. roseus* seedlings in a gyrotary shaker operated at 80 rpm, 24° ± 2 °C under a 16 h day light cycle and subcultured approximately every week. A 2.5 g fresh weight inoculum typically produced 29 g fresh weight after 35 d growth.

There are several interesting reports describing the correlation between differentiation and production of secondary metabolites. Hiraoka and Tabata [113] successively transferred the callus of *Datula meteloides* from a medium with auxins to one without and then cultivated continuously. As seen in Table 9, shoots, stems and roots were differentiated by turn and tropane alkaloid levels increased. In *Digitalis purpurea* cultures, Hagimori et al. [114] showed stimulation of digitalis cardenolide production by organ redifferentiation calli. A similar phenomena was also found in rotenone formation using *Derris elliptica* [115] and morphinane alkaloid production using *Papaver somniferum* [116]. Furuya et al. [116] investigated the correlation between the stage of morphological differentiation and producing ability of the alkaloids using *P. somniferum*, and found a green callus which differentiates epidermis or vascular bundles produced the alkaloids. A limited degree of tissue differentiation occurred and the cell contained codeine as a main alkaloid while the level of morphine increased as differentiation progressed.

Table 9. Relationship between producing ability of alkaloids and differentiation (Hiraoka et al. [113])

Plants	Alkaloid concentration (% dry wt.)
Callus	1×10^{-2}
Shoots-forming callus	1.5×10^{-2}
Growing shoots	2×10^{-2}
Roots-forming shoots	3×10^{-2}
Leaf of young plant	1×10^{-1}
Leaf of matured plant	3×10^{-1}

Kamo et al. [117] analyzed the producing capability of calli, derived meristemoids, redifferentiated roots and redifferentiated shoots of *P. somniferum*. As seen in Table 10, the callus tissue produced both thebaine and codeine but derived meristemoids accumulated only thebaine. On the other hand, redifferentiated shoots produced morphine, thebaine and codeine. Root cultures derived from cell suspension cultures of *P. bracteatum* were shown to produce thebaine in a yield of 0.03%, by Zito and Staba [118]. They [79] also reported that axenic callus and shoot cultures of *Pyrethrum cinerarifolium* had an ability to produce pyrethrin. They isolated a few high yielding strains which were derived from high yielding plant selections. One isolate accumu-

Table 10. Production of alkaloids by *Papaver somniferum* calli and redifferentiated tissues (Kamo et al. [117])

Tissue	Total	Alkaloid (%)		
	(μg g^{-1} dry wt)	Thebaine	Codeine	Morphine
Intact seedling				
Shoot	1040	12	2	86
Root	999	43	6	51
Callus-induced organs				
Meristemoids, removed from callus (no buds)	13.2	100	0	0
Cultured meristemoids with buds	Trace	100	0	0
Shoots	187	40	12	48

lated 11.3 mg total pyrethrines per 100 g dry weight but it subsequently differentiated into a shoot culture following the first analysis. Therefore, Zito and Staba [118] concluded that differentiated cultures tended to produce more pyrethrines than did callus cultures. Using an established shoot culture derived from the disc floret of the plant, 341.8 mg of pyrethrines per 100 g wt., were obtained.

Ozeki and Komamine [119] have recently presented interesting results on the relationship between differentiation and anthocyanin production using *Daucus carota* cells. The cells were fractionated by Ficoll density gradient centrifugation. In the density fraction (> 14% of Ficoll) somatic embryos were formed in a medium containing 10^{-7} zeatin but anthocyanin was scarcely produced. On the other hand, the cells in the lower density fraction (< 12% of Ficoll) synthesized anthocyanin in the same medium but formed few embryos. 40 to 50% of the total cells in the higher cell fraction synthesized anthocyanin at a maximum.

A shoot culture established from stem nodes of *Dioscorea composita* was shown to produce diosgenin in a yield similar to in vivo synthesis [120].

Lindsey and Yeoman [121] have investigated the relationship between alkaloid content, growth rate and differentiation in seven solanaceous species in culture. They observed an inverse correlation between the growth phase on the one hand and alkaloid content and cell organization (aggregation) on the other.

Thus, it seems that in many cases morphological differentiation may be necessary to obtain higher yields of secondary metabolites. However, this is not always desirable in large scale cultivation because, generally the culture period for differentiation tissue is rather longer than that of undifferentiated cells. It is obvious that a shorter culture period is required to avoid microbial contamination and to lower the production cost.

4 Products of Interest to Industry

Among a variety of products reported by many research groups [122] as present in cell cultures, some of them are very interesting because of their high added value. These compounds are unique metabolites in higher plants and cannot be produced by either chemical synthesis or a microbial process. However, plant cell culture is still not sufficiently developed to compete with these technologies. Nevertheless several interesting examples from an industrial point of view can now be described.

4.1 Alkaloids

A variety of alkaloids have been used as pharmaceuticals and most of them are plant metabolites. The typical tropane alkaloids, atropine, hyocyamine, scopolamine and cocaine, are widely used as blockers of the parasympathetic nervous system such as anodyne and antispasmodic.

Research on production of these useful alkaloids has been carried out for more than 20 years. However, industrial production has not yet succeeded because of the low producing ability of the cultured cells. Plants used for these studies are mainly *Atropa belladonna*, *Hyocyamus niger*, *Datura meteloides* and others, and various approaches to increase their producing ability have been tried by many researchers.

Recently, Mitsuno et al. [123] selected a high tropane alkaloid producing strain of *Hyocyamus niger* which produced about 7 times more hyocyamine, (13.9×10^{-3}% fresh cells), than that of the parent strain. According to their results, there was no direct relationship between high producing ability and variation of chromosomal numbers. Smordin et al. [124] with medium containing a high concentration of sucrose (sufficient to provide 15 atmospheric osmotic pressure) obtained a higher titer of atropine. They suggested that the osmotic pressure played a role in increasing product formation.

Nicotine is a representative alkaloid in tobacco which belongs to the pyridine and piperidine alkaloid group. It has a strong insecticidal effect but it does not make any sense to manufacture this by tissue culture since nicotine is available inexpensively from waste leaves of tobacco plants. However, a tobacco cell is one of the easiest cells to cultivate and therefore many papers dealing with the cell have been published. As a result of cell cloning techniques using *N. rustica*, Tabata et al. [125] isolated a very stable cell line showing 0.29% nicotine production even after 54 generations. According to a patent filed by Smith [126], 152 mg of nicotine per litre of the medium containing 0.2 mg l^{-1} NAA and 0.02 mg l^{-1} kinetin was accumulated after 24 d cultivation an *N. tabacum* culture.

It should be noted that until 10 years ago, tissue culture lines produced only small amounts of nicotine, and this fact was used by the Japan Tobacco and Salt Public Corporation as a basis for the production of "nicotine-less tobacco".

It is well-known that ajmalicine and reserpine have been used as drugs for hypertension and circulatory diseases. The excellent research carried out by Zenk and his group [59] in production of high levels of ajmalicine and serpentine using radioimmunoassay have undoubtedly encouraged many workers, studying secondary metabolite production, as already mentioned.

Among more than 100 alkaloids found in *Catharanthus roseus*, vincristine and vinblastine are the most attractive targets since they are already on the market as potent antitumor drugs. Although many efforts to produce them by cell culture have been made by several research groups, their production, unfortunately, still depends upon extraction from intact plants. Kurz et al. [127] reported the accumulation of catharanthine using a selected cell line of *C. roseus* and they also [128] recognized biotransformation of vindoline and catharanthine to 3′,4′-dehydrovinblastine and that of the latter compound to leurosine and catharanthine. These biotransformations are of interest if the substrates could be supplied inexpensively. Recently, Hirata et al. [129] reported a direct production of vinblastine from sucrose and inorganic nitrogen sources using a *C. roseus* suspension culture, in which the maximum level was 5.2 μg of vinblastine per g of the dried cells when the cells were cultivated in Murashige and Skoog (MS) medium containing 1.0 mg l^{-1} benzyladenine [130]. A strain which produced 10 times higher levels of the alkaloid than the parent was obtained by a cell cloning technique but the productivity was unstable.

Since berberine has a big market in the Orient, several researchers have been working on its production, as described previously. It would be of interest to apply plant tissue culture technology to the production of berberine if the price were higher.

Some alkaloids have potent antineoplastic activity [131] and such types of compound are desirable products because of their added value. Camptothecin found by Wall et al. [132] was a target drug about 10 years ago since it was shown to have potential antitumor activity against the mouse leukemia L-1210 and the Walker 256 rat carcinosarcoma. Accumulation of the alkaloid was recognized by Misawa and his group in cultured *Camptotheca acuminata* cells in 1974 [133]. Unfortunately this compound was dropped from the NCI screening program of anticancer compounds after clinical trials indicated it to be toxic.

Cephalotoxin esters such as homoharringtonine and harringtonine found in *Cephalotaxus harringtonia* [134] are also antineoplastic alkaloids. Delfel [135] reported its accumulation at 1 to 3% of the level in the intact plant and the author and his colleagues established a suspension culture of *C. harringtonia* [136] and radioimmunoassay to determine homoharringtonine [137]. The cells were cultivated in MS liquid medium containing 1% sucrose and 3% NAA at 25 °C on a rotary shaker for 15 d. According to a modified method of Powells', a crude alkaloid was obtained and then

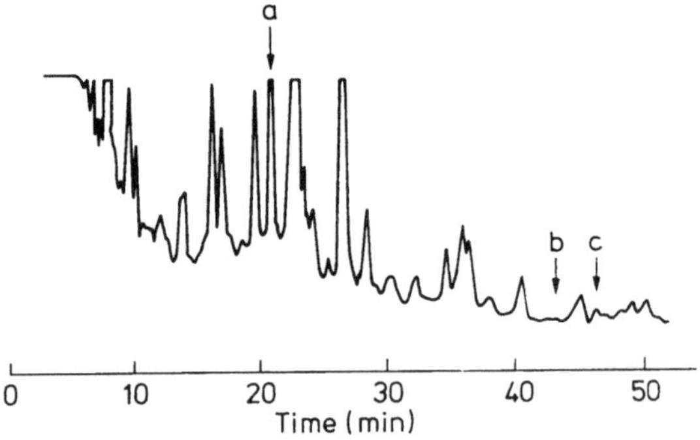

Fig. 13. Gas chromatography of crude alkaloid fraction obtained from suspension cultured cells of *Cephalotaxus harringtonia* [136].
Silicon OV-1 on Gaschrom. Q; Glass column (1 m); Temperature: 150 to 300 °C at 2 °C per min; H_2: 0.6 kg cm^{-2}; Detector FID.
a: Methyl lignocerate (C_{24}: 0), internal standard;
b: Harringtonine;
c: Homoharringtonine

trimethylsilylated with bis(trimethylsilyl)-trifluoroacetamide for gas chromatography. As seen in Fig. 13, trace amounts of both homoharringtonine and harringtonine were detected in the culture cells, which corresponded to approximately 1.7×10^{-5}% as homoharringtonine in the dried cells from their cytotoxicity against KB cells [136]. The level was 1/50 to 1/100 of that in the intact plant.

4.2 Steroids

Since several kinds of cardenolides and steroid hormones are on the market, research on the production of these compounds from plant cell cultures is also active. Diosgenin, which is a raw material for steroidal hormones, is a good target because of its restricted supply from *Dioscorea* plants. It was reported that *Dioscorea deltoidea* cells could accumulate 2.5% diosgenin dry weight if growth on media containing cholesterol and 2,4-D [138].

As I mentioned previously, industrial biotransformation of β-methyldigitoxin has been expected for the last 10 years. However, it seems that there are economical problems in producing β-methyldigoxin in industry in spite of the many efforts of Reinhard et al.

4.3 Terpenoids

Antitumor compounds, tripdiolide and triptolide isolated by Kupchan and his group in 1972 [139] are also promising compounds in the NCI screening program. The production of these diterpenetriepoxide compounds were independently studied by Kutney and his group at the University of British Columbia, Canada [140] and Misawa and his group at Kyowa Hakko Kogyo Co., Ltd., Japan [136, 141] under a research contract from NIH. Tripdiolide was found to be accumulated in *Tripterygium wilfordii* suspension cultured cells in about 10 times higher amounts than in the intact plant. A typical time course of the cell growth and the production of tripdiolide is shown in Fig. 14 [141]. The cells also produced triptolide. As seen in this figure, 1 mg l^{-1}

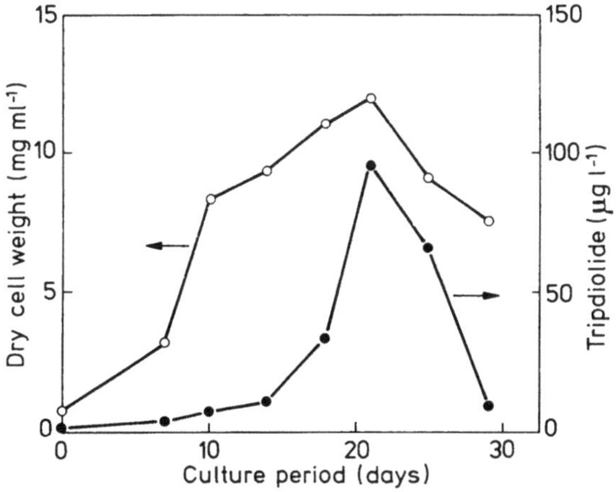

Fig. 14. A time course of tripdiolide production by *Tripterygium wilfordii* cell suspension cultures [141]

kinetin and 0.1 mg l^{-1} NAA were employed as phytohormones and the intracellular level of tripdiolide reached a maximum of 95 μg in a litre of culture broth, on the 21st day of the cultivation. The relationship between growth and production indicated that this is a typical secondary metabolite production which is similar to microbial production of secondary metabolites. Although the level of tripdiolide in the culture filtrate has not been determined, the existence of a high level of the compound in the filtrate was presumed by a rough estimation.

Cells of *Panax ginseng* containing high levels of ginsenoside Rb_1 and Rg_1 have been cultivated in tanks, and the industrial production of dried cells is being discussed in Japan [142]. Furuya et al. [143] isolated a cell line of *P. ginseng* having high yields of saponins and cell mass and cultivated this in a 30 l reactor. The highest content in the cells was 57.2 mg of total saponins per litre of the medium at 28 days cultivation.

4.4 Quinones

Production of ubiquinone-10, which has a big market as a drug for heart diseases in Japan, has been studied by the Japan Tobacco and Salt Public Corporation using large fermentors. However, the production cost is still higher than that by microbial process. If the compound could be obtained as a by-product of tobacco cells, a raw material of cigarettes, it might be interesting economically. In contrast Mitsui Petrochemical Co., Ltd., in Japan recently announced the commercial production of shikonin derivatives for a pigment in lipsticks. Although the total amount produced is not large, it is the first case of commercial production of a secondary metabolite using plant cell cultures.

There is a large number of papers dealing with production of some pharmacologically active quinones such as emodin, aloe-emodin, rhein etc., using *Rheum palmatum*, *Cassia obtusifolia*, *Cassia angustifolia* and other callus tissues, but they are still far from industrial application.

4.5 Miscellaneous

Proteinase inhibitors isolated from *Scopolia japonica* [22,144] cultured cells and plant virus inhibitors isolated from *Phytolacca americana* [16] and *Agrostemma githago* [17] cultured cells are very interesting polypeptides. However, the commercial development of these compounds as new human and agricultural drugs was suspended because of the high cost and long periods of time required to get approval from the Ministry of Health and Welfare. The existence of these interesting substances in cultured cells was recognized during the course of an extensive screening program at Kyowa Hakko. The finding of new types of useful compounds in the cultured cell extracts is another approach in the study of plant cell culture. Recently, a group in W. Germany started this sort of research to look for new biologically active compounds [145].

Determination of the structure of such compounds might provide some new ideas for drug design, particularly if the compounds had novel skeletons. The compound having a new type of skeleton which relates to biological activity might give some idea for drug design unless the compound can be produced by large scale cultivation of plant cells economically.

Several antibiotic substances were found in callus tissues by several researchers [146] but the activity of these is not high enough to be applicable to human diseases.

Maytansine was decribed by Kupchan et al. in 1972 [139] as a high potency antineoplastic compound in Maytenus and Putterlickia plants. Since this compound is an anthamacrolide compound and its concentration in the trees is very low, it was thought that maytansine was a product of microorganisms infecting the trees. From the callus tissue of *Putterlickia verrucosa*, the author and his colleagues isolated a trace amount of cytotoxic compound, 15 µg in 300 g of the dried callus, which was very similar to maytansine as far as they examined by HPLC, TLC and UV spectra [136]. They also recognized a cytotoxic compound which was very similar to baccharin, and trichothecene, in the callus of *Baccharis megapotamica*.

It is difficult to produce pigments for food by cell cultures economically, since the cost of these materials is low. For example, a large amount of betanin was accumulated in the cultured cells of *Phytolacca americana* [148], *Beta vulgaris* [149], *Chenopodium centrorubrum* [149] etc., but it is very hard to manufacture this more cheaply than by extraction from peels of *Beta vulgaris* which are waste products in the sugar industry.

Production of anthocyanin is also another target as a pigment [119, 150–152]. Mitsui Petrochemical Co., Ltd., filed a patent relating accumulation of anthocyanins in *Derris* sp. [153], and Minoda et al. described their production using *Vitis* suspension cultures in their patent [154], in which 161 mg of the pigments were available in 500 ml of medium after 17d cultivation. Using carrot cell cultures, Dougall et al. [150, 151] have worked extensively on anthocyanin production. They could isolate a high producing cell line. The differences in the productivity between high and low density cells found by Ozeki et al. [119] were mentioned previously. The effect of auxins on the production was also examined by Smith [152] using *Strobilonthes dyeriana*. Unfortunately, the production of anthocyanins by cell culture is also too costly to realize commercially.

Stevioside is a sweetener in the leaves of *Stevia rebaudiana* which is commercially available. The compound is a glucoside similar to saponin and is 300 times sweeter than sucrose. Kibun Co., Ltd., in Japan filed a patent on its production by *S. rebaudiana* cell cultures [155] but the yield was not clear. There is another patent described by Chugai Pharmaceutical Co., in Japan [156], which showed production of stevioside from 5 mg steviol in 60% conversion yield using 20 g of the callus. A group in Unilever Co. [157], cloned a gene for thaumatin which is a proteinaceous sweetener from the *Thaumatococcus daniellii*, a plant indigenous to tropical West Africa and incorporated it into *E. coli*. Although this technique has not yet been applied in its commercial production, recombinant DNA technology will become a useful method to produce plant metabolites.

5 Conclusion

With the advance of biotechnology, research on plant tissue cultures is becoming more active not only in academic institutions but also in industry. From an economical point of view, there are at least three applications of this field; plant breeding, useful metabolite production and micropropagation. Except for micropropagation through tissue culture, it has been thought that the two other objectives take too long a time to contribute to industry. However, as seen in shikonin derivative production, the industrial application of this technology to manufacture metabolites seems likely to be realized in the near future although there are still some problems.

In order to overcome these problems, it is obvious that more extensive fundamental research will be needed collaborating with a number of researchers in other scientific fields. Furthermore, it should be noted that the selection of the most suitable products for plant cell culture is very important as the author has emphasized here repeatedly.

6 Acknowledgements

I am very grateful to all the members of the Research Institute for Biological Sciences at Kyowa Hakko Kogyo Co., Ltd., in Japan for their helpful comments, as well as to Dr. Derek Burke at Allelix Inc., in Canada for his kind help in the preparation of the English version of this manuscript.

7 References

1. Zenk, M. H.: Frontiers of Plant Tissue Culture 1978 (Thorpe, T. A., ed.), Int. Assoc. for Plant Tissue Culture, p. 1. Univ. of Calgary, 1978
2. Kurz, W. G. M., Constabel, F.: Adv. Appl. Microbiol. *25*, 209 (1979)
3. Fowler, M.: Progress in Industrial Microbiol. *16*, 207 (1982)
4. Staba, E. J. (ed.): Plant Tissue Culture as a Source of Biochemicals, CRC Press Inc., Florida 1980
5. Staba, E. J.: Plant Tissue Culture 1982, Proc. 5th Intl. Cong. Plant Tissue & Cell Culture (Fujiwara, A., ed.), p. 25. Maruzen Co. Ltd., Tokyo 1982
6. Tulecke, W., Nickell, L. G.: Science *130*, 863 (1959)
7. Mandels, M.: Adv. Biochem. Eng. *2*, 201 (1972)
8. Street, H. E. (ed.): Plant Tissue and Cell Culture, Blackwell Scientific Publ., London 1973
9. Martin, S. M.: Plant Tissue Culture as a Source of Biochemicals (Staba, E. J., ed.), p. 149. CRC Press Inc., Florida 167, 1980
10. Misawa, M.: Plant Tissue Culture and Its Bio-technological Application (Barz, W., Reinhard, E., Zenk, M. H., eds.), p. 17. Springer Berlin, Heidelberg, New York 1977
11. Barz, W., Reinhard, E., Zenk, M. H. (eds.): Plant Tissue Culture and Its Bio-technological Application. Springer Berlin, Heidelberg, New York 1977
12. Misawa, M., Suzuki, T.: Appl. Biochem. and Biotechnol. *7*, 205 (1982)
13. Goldstein, W. F., Lasure, L. L., Ingle, M. B.: Plant Tissue Culture as a source of Biochemicals (Staba, E. J., ed.), p. 191. CRC Press Inc., Florida 1980
14. Zenk, M. H., El-Shagi, H., Arens, H., Stöckigt, J., Weiler, E. W., Deus, B.: Plant Tissue Culture and Its Bio-technological Application (Barz, W., Reinhard, E., Zenk, M. H., eds.), p. 27. Springer Berlin, Heidelberg 1977
15. Zenk, M. H., El-Shagi, H., Ulbrich, B.: Naturwissenschaften *64*, 585 (1977)
16. Misawa, M., Hayashi, M., Tanaka, H., Ko, K., Misato, T.: Biotech. Bioeng. *17*, 1335 (1975)
17. Takayama, S., Misawa, M., Ko, K., Misato, M.: Physiol. Plant. *41*, 313 (1977)
18. Tanaka, H., Machida, Y., Mukai, N., Misawa, M.: Agric. Biol. Chem. *38*, 987 (1974)
19. Matsumoto, T., Okunishi, K., Nishida, K., Noguchi, M., Tamaki, E.: ibid. *35*, 543 (1971)
20. Noguchi, M., Matsumoto, T., Hirata, Y., Yamamoto, K., Katsuyama, A., Kato, A., Azechi, S., Kato, K.: Plant Tissue Culture and Its Biotechnological Application (Barz, W., et al. eds.), p. 85, 1977
21. Hashimoto, T., Azechi, S., Sugita, S., Suzuki, K.: Proc. 5th Intl. Cong. Plant Tissue and Cell Culture (Fujiwara, A., ed.), p. 403. Maruzen Co., Ltd., Tokyo 1982
22. Misawa, M., Tanaka, H., Chiyo, O., Mukai, N.: Biotech. Bioeng. *17*, 305 (1975)

23. Kato, K., Matsumoto, T., Koiwai, A., Mizukami, S., Nishida, K., Noguchi, M., Takami, E.: Fermentation Technol. Today (Terui, G., ed.), p. 689. Kyoto Soc. Fermentation Technol. 1972
24. Sasse, F., Knobloch, K.-H., Berlin, J.: Proc. 5th Intl. Cong. Plant Tissue and Cell Culture (Fujiwara, A., ed.), p. 343. Maruzen Co., Ltd., Tokyo 1982
25. Hagimori, M., Matsumoto, T., Obi, Y.: ibid. p. 349 (1982)
26. Takayama, S., Misawa, M., Ko, K., Misato, T.: Physiol. Plant *41*, 313 (1977)
27. Hayashi, M., Takayama, S., Misawa, M.: in preparation
28. Brain, K. R.: Plant Sci. Letters *7*, 157 (1976)
29. Ikeda, T., Matsumoto, T., Noguchi, M.: Phytochem. *15*, 568 (1976)
30. Staba, E. J., Kaul, B.: U.S. Pat. 3,628,287 (1971)
31. Ambid, C., Roustan, J.-P., Nef, C., Fallot, J.: Proceedings of 5th IAPTC (Fujiware, A., ed.), p. 331. Maruzen Co., Ltd., Tokyo 1982
32. Okazaki, M., Hino, F., Kominami, K., Miura, Y.: Agric. Biol. Chem. *46*, 601, 609 (1982)
33. Ikuta, A., Itokawa, H.: Phytochem. *21*, 1419 (1982)
34. Takahashi, S., Shudo, K., Okamoto, T., Yamada, K., Isogai, Y.: ibid. *17*, 1201 (1978)
35. Fujita, Y., Hara, Y., Suga, C., Morimoto, T.: Plant Cell Reports *1*, 61 (1981)
36. Fujita, Y., Maeda, Y., Tan, H., Suga, C.: Oral Presentation at 8th Symp. of Plant Tissue Culture in Japan. Toyama, July 1983
37. Mitsui Petro Chemical Co., Ltd., Japan Patent (Open) 1982-39778 (1982)
38. Wagner, F., Vogelmann, H.: Plant Tissue Culture and Its Biotechnological Application (Barz, W., et al., eds.), p. 245. Springer Berlin, Heidelberg 1977
40. Alfermann, A. W., Reinhard, E.: Proc. 5th Intl. Cong. Plant Tissue and Cell Culture (Fujiwara, A., ed.), p. 401. Maruzen Co. Ltd., Tokyo 1982
41. Smart, N. J., Morris, P., Fowler, M. W.: ibid., 397 (1982)
42. Tal, B., Goldberg, I.: Planta medica *44*, 107 (1982)
43. Hagimori, M., Matsumoto, T., Obi, Y.: Plant Physiol. *69*, 653 (1982)
44. DiCosmo, F., Towers, G. H. N.: Phytochemical Adaptations to Stress (Timmermann, B. N., Steelink, C., Loewus, F. A., eds.) p. 91–175. Plenum Publishing Corp. (1984)
45. Chan, W., Staba, E. J.: Lloydia *28*, 55 (1965)
46. Tabata, M., Yamamoto, H., Hirakawa, N.: In: Les Cultures de Tissue de Plantes, p. 389. CNRS, Paris 1971
47. Alfermann, A. W., Reinhard, E.: Bulletin de la Société Chemique de France II-35–45 (1980)
48. Reinhard, E.: Tissue Culture and Plant Science 1974 (Street, H. E., ed.), p. 433. Academic Press 1974
49. Alfermann, S. W., Reinhard, E.: Proc. 5th Intl. Cong. Plant Tissue and Cell Culture (Fujiwara, A., ed.), p. 401. Maruzen Co. Ltd., Zokyo 1982
50. Alfermann, S. W.: Personal Communication
51. Kurz, W. G. W., Constabel, F., Kutney, J. P.: Proc. 5th Intl. Cong. Plant Tissue and Cell Culture (Fujiwara, A., ed.), p. 361. Maruzen Co. Ltd., Tokyo 1982
52. Tam, W. H. J., Kurz, W. G. W., Constabel, F., Chatson, K. B.: Phytochem. *21*, 253 (1982)
53. Jones, A., Veliky, I. A.: Planta medica *42*, 160 (1981)
54. Lui, J. H. C., Staba, E. J.: Phytochem. *18*, 1913 (1979)
55. Hiraoka, N.: J. Nat. Products *44*, 285 (1981)
56. Umetani, Y., Tanaka, S., Yamaura, T., Yamanaka, K., Tabata, M.: 8th Symp. Plant Tissue Culture, Toyama, Japan, July 1983
57. Jean, F., Michael, M., Christian, A.: Proc. 5th Intl. Cong. Plant Tissue and Cell Culture (Fujiwara, A., ed.), p. 389. Maruzen Co. Ltd., Tokyo 1982
58. Mangold, H. K., Radwan, S. S.: Plant Cell Cultures: Results and Prospectives (Sala, F., et al., eds.), p. 363. Elsevier and North-Holland Biomedical Press 1980
59. Zenk, M. H., El-Shagi, H., Arens, H., Stöckigt, J., Weiler, E. W., Deus, B.: Plant Tissue Culture and Its Bio-technological Application (Barz, W., et al., eds.), p. 27. Springer Berlin, Heidelberg 1977
60. Yamakawa, T., Ohtsuka, H., Onomichi, K., Kodama, T., Minoda, Y.: Proc. 5th Intl. Cong. Plant Tissue and Cell Culture (Fujiwara, A., ed.), p. 273. Maruzen Co. Ltd., Tokyo 1982
61. Raveh, D., Huberman, F., Galun, E.: In Vitro *9*, 216 (1973)
62. Yamamoto, Y., Mizuguchi, R., Yamada, Y.: Theor. Appl. Genet. *61*, 113 (1982)

63. Constabel, F., Kurz, W. G. W., Kutney, J. P.: Proc. 5th Intl. Cong. Plant Tissue and Cell Culture (Fujiwara, A., ed.), p. 301. Maruzen Co., Ltd., Tokyo 1982
64. Constabel, F.: 1st Intl. Symp. on Life Sci. '82, Ohtsu, Japan 1982
65. Yamada, Y., Sato, F.: Phytochem. *20*, 545 (1981)
66. Sato, S., Endo, T., Hashimoto, T., Yamada, Y.: Proc. 5th Intl. Cong. Plant Tissue and Cell Culture (Fujiwara, A., ed.), p. 319. Maruzen Co., Ltd., Tokyo 1982
67. Furuya, T., Synono, K., Ikuta, A.: Phytochem. *11*, 175 (1972)
68. Fukui, H., Nakagawa, K., Tsuda, S., Tabata, M.: Proc. 5th Intl. Cong. Plant Tissue and Cell Culture (Fujiwara, A., ed.), p. 313. Maruzen Co. Ltd., Tokyo 1982
69. Khanna, P., Sharma, O. P., Saluja, M.: ibid. p. 311 (1982)
70. Ikuta, A., Itokawa, H.: Phytochem. *21*, 1419 (1982)
71. Tabata, M., Mizukami, H., Hiraoka, N., Konoshima, M.: ibid. *13*, 927 (1974)
72. Tabata, M.: Plant Tissue Culture and Its Bio-technological Application (Barz, W., et al., eds.), p. 3. Springer Berlin, Heidelberg 1977
73. Watanabe, K., Yano, S., Yamada, Y.: Phytochem. *21*, 513 (1982)
74. Watanabe, K., Yamada, Y.: Proc. 5th Intl. Cong. Plant Tissue and Cell Culture (Fujiwara, A., ed.), p. 357. Maruzen Co., Ltd., Tokyo 1982
75. Matsumoto, T., Ikeda, T., Kanno, N., Kisaki, T., Noguchi, M.: Agric. Biol. Chem. *44*, 967 (1980)
76. Matsumoto, T., Ikeda, T., Kanno, N., Obi, Y., Kisaki, T., Noguchi, M.: ibid. *45*, 1627 (1981)
77. Matsumoto, T., Ikeda, T., Okimura, C., Obi, Y., Kisaki, T., Noguchi, M.: Proc. 5th Intl. Cong. Plant Tissue and Cell Culture (Fujiwara, A., ed.), p. 275. Maruzen Co., Ltd., Tokyo 1982
78. Ikeda, T., Matsumoto, T., Noguchi, M.: Agric. Biol. Chem. *40*, 1765 (1976)
79. Zieg, P. G., Zito, S. W., Staba, E. J.: Planta medica *48*, 88 (1983)
80. Deus, B., Zenk, M. H.: Biotech. Bioeng. *24*, 1965 (1982)
81. Böhm, H.: Proc. 5th Intl. Cong. Plant Tissue and Cell Culture (Fujiwara, A., ed.), p. 325. Maruzen Co., Ltd., Tokyo 1982
82. Berlin, J., Sasse, F., Knobloch, K. H.: ibid. p. 329 (1982)
83. Widholm, J. M.: Plant Tissue Culture as a Source of Biochemicals (Staba, E. J., ed.), p. 99. CRC Press Inc., Florida 1980
84. Scott, A. J., Mizukami, H., Lee, S. L.: Phytochem. *18*, 795 (1979)
85. Schallenberg, J., Berlin, J.: Z. Naturforsch. *34C*, 541 (1979)
86. Berlin, J.: Z. Pflanzenphysiol. *97*, 317 (1980)
87. Berlin, J., Knoblock, K. H., Höfle, G., Witte, L.: J. Nat. Prod. *45*, 83 (1982)
88. Berlin, J.: Personal Communication
89. Nishi, A., Yoshida, A., Moir. M., Sugano, N.: Phytochem. *13*, 1653 (1974)
90. Sung, Z. R.: Genetics *84*, 51 (1976)
91. Müller, A. J., Grafe, R.: Int. Bot. Congr. Abst. *12*, 304 (1975)
92. Schieder, O.: Mol. Gen. Genet. *149*, 251 (1976)
93. Widholm, J. M.: Molecular Genetic Modification of Eucaryotes (Rubinstein, I., et al., eds.), p. 57. Academic Press, New York 1977
94. Guang-zhi, Z., Jing-bo, H., Shi-ling, W.: Proc. 5th Intl. Cong. Plant Tissue and Cell Culture (Fujiwara, A., ed.), p. 339. Maruzen Co., Ltd., Tokyo 1982
95. Deus, B.: Proc. Intl. Symp. Plant Cell Culture (Alfermann, A. W., Reinhard, E., eds.), p. 118. Ges. S.U. mbH., Munich 1978
96. Negrutin, I., Dirks, R.: Proc. 5th Intl. Cong. Plant Tissue and Cell Culture (Fujiwara, A., ed.), p. 461. Maruzen Co., Ltd., Tokyo 1982
97. Gebhardt, C., Fankhauser, H., King, P. J.: ibid. p. 463 (1982)
98. Shimamoto, K., Aeschbacher, G., King, P. J.: ibid. p. 465 (1982)
99. Christian, T. H., Oertli, J. J.: ibid. p. 467 (1982)
100. Ho, C., Loo, S., Xu, Z., Xu, S., Li, W.: ibid. p. 469 (1982)
101. Berlin, J.: Personal Communication
102. Franke, J., Böhm, H.: Biochem. Phys. Pflanzen. *170*, 501 (1982)
103. Yamakawa, T., Kobayashi, E., Shinagawa, R., Kodama, T., Minoda, Y.: Oral presentation at 8th Meeting of Plant Tissue Culture in Japan, Toyama, Japan July 1983
104. Brodelius, P., Mosbach, K.: Adv. Appl. Microbiol. 1982 (Laskin, A. I., ed.), Vol. 28, p. 1. Academic Press, New York

105. Brodelius, P., Mosbach, K.: J. Chem. Tech. Biotechnol. *32*, 330 (1982)
106. Brodelius, P., Deus, B., Mosbach, K., Zenk, M. H.: FEBS Lett. *103*, 93 (1979)
107. Brodelius, P., Linse, L., Nilsson, K.: Proc. 5th Intl. Cong. Plant Tissue and Cell Culture (Fujiwara, A., ed.), p. 371. Maruzen Co., Ltd., Tokyo 1982
108. Alfermann, A. W., Reinhard, E.: ibid. p. 401 (1982)
109. Veliky, I. A., Jones, A.: Biotechnol. Letters *10*, 551 (1981)
110. Yoshikawa, T., Tahira, M., Furuya, T.: Oral Presentation at 8th Meeting of Plant Tissue Culture in Japan, Toyama, Japan, July 1983
111. Wiermann, R.: The Biochemistry of Plants, p. 85. Academic Press Inc. 1981
112. Krueger, R. J., Carew, D. P., Lui, H.-C., Staba, E. J.: Planta medica *45*, 56 (1982)
113. Hiraoka, N., Tabata, N.: Phytochem. *13*, 1671 (1974)
114. Hagimori, M., Matsumoto, T., Kisaki, T.: Plant and Cell Physiol. *21*, 1391 (1980)
115. Kodama, T., Yamakawa, T., Minoda, Y.: Agric. Biol. Chem. *44*, 2387 (1980)
116. Furuya, T., Yoshikawa, T.: Oral Presentation at 8th Plant Tissue Culture, Toyama, Japan, July 1983
117. Kamo, K. K., Kimoto, K. W., Hsu, A. F., Mahlberg, P. G., Bills, D. D.: Phytochem. *21*, 219 (1982)
118. Zito, S. W., Staba, E. J.: Planta medica *45*. 53 (1982)
119. Ozeki, Y., Komamine, A.: Physiol. Plant. *53*, 570 (1981)
120. Heble, M. R., Staba, E. J.: Planta medica Suppl. 120 (1980)
121. Lindsey, K., Yeoman, M. M.: J. Exp. Botany *34*, 1055 (1983)
122. Nickell, L. G.: Plant Tissue Culture as a Source of Chemicals (Staba, E. J., ed.), p. 236. CRC Press, Florida 1980
123. Mitsuno, M., Hashimoto, T., Yamada, Y.: Oral Presentation at the 8th Meeting of Plant Tissue Culture, Toyama, Japan, July 1983
124. Smordin, V. V., Smordin, A. V., Bereznegovskaya, L. N.: Rust Resur. *9*, 550 (1973)
125. Tabata, M., Hiraoka, N.: Physiol. Plant. *38*, 19 (1976)
126. Smith, H.: U.K. Pat. Appl. 2,025,952 (1980)
127. Kurz, W. G. W., et al.: Planta medica *42*, 22 (1981)
128. Kutney, J. P.: Pure and Appl. Chem. *54*, 2523 (1982)
129. Hirata, M., Okazaki, M., Miura, Y.: Oral Presentation of the 8th Meeting of Plant Tissue Culture, Toyama, Japan, July 1983
130. Hirata, M., Inoue, T., Miura, Y., Okazaki, M.: Oral Presentation at the 35th Ann. Meeting of Fermentation Technology in Japan, Osaka, Japan, Nov. 1983
131. Douros, J., Suffness, M., Carter, S. K., Sakurai, Y. (eds.): Recent Results in Cancer Research, Springer-Verlag, Berlin, Heidelberg, p. 21, 1980
132. Wall, M. E., Wani, M. C., Cook, C. E., Palmer, K. H., McPhail, A. T., Sun, G. A.: J. Amer. Chem. Soc. *88*, 3888 (1966)
133. Sakato, K., Tanaka, H., Mukai, N., Misawa, M.: Agric. Biol. Chem. *38*, 217 (1974)
134. Powell, R. G., Weisleder, D., Smith, C. R., et al.: Tetrohedron Lett. *1970*, 815 (1970)
135. Delfel, N. E., Rothfus, J. A.: Phytochem. *16*, 1595 (1977)
136. Misawa, M., Hayashi, M., Takayama, S.: Planta medica *49*, 115 (1983)
137. Takayama, S., et al.: In Preparation
138. Kaul, B., Stohs, S. J., Staba, E. J.: Lloydia *32*, 347 (1969)
139. Kupchan, S. M., Court, W. A., Dailey Jr., R. G., et al.: J. Am. Chem. Soc. *94*, 7194 (1972)
140. Kutney, J. P. et al.: Planta medica *48*, 158 (1983)
141. Hayashi, M., Takayama, S., Misawa, M.: Proc. 5th Intl. Cong. Plant Tissue and Cell Culture (Fujiwara, A., ed.), p. 291. Maruzen Co., Ltd., Tokyo, Japan 1982
142. Furuya, T.: ibid. p. 269 (1982)
143. Furuya, T., Yoshikawa, T., Orihara, Y., Oda, H.: J. Nat. Products *47*, 70 (1984)
144. Sakato, K., Tanaka, H., Misawa, M.: Eur. J. Biochem. *55*, 211 (1975)
145. Berlin, J.: Personal Communication
146. Kurosaki, F., Nishi, A.: Phytochem. *22*, 669 (1983)
147. Kupchan, S. M., Komoda, Y., Court, W. A., et al.: J. Am. Chem. Soc. *95*, 1354 (1972)
148. Misawa, M., Hayashi, M., Nagano, Y., Kawamoto, T.: Japan Patent (Kokai) 73-6153 (1973)
149. Komatsu, K., Nozaki, W., Takemura, M., Umemori, S., Nakaminami, M.: Japan Patent (Kokai) 74-85286 (1974)

150. Dougall, D. K.: Biotech. Bioeng. *22*, 337 (1980)
151. Dougall, D. K.: Planta *149*, 292 (1980)
152. Smith, S. L., et al.: J. Nat. Product *44*, 605 (1981)
153. Mitsui Petrochemicals Co., Ltd. Japan Patent (Kokai) 79-11281 (1979)
154. Minoda, Y., et al.: ibid. 80-118392 (1980)
155. Kibun Co., Lrd., ibid. 76-19169 (1976)
156. Chugai Pharmaceutical Co. Ltd., ibid. 80-19009 (1980)
157. Edens, L., Heslinga, L., Klok, R., Ledeboer, A. M., Maat, J., Toonen, M., Visser, C., Verrips, C. T., Leeuwenhoek, Antonie: J. Microbiol. *48*, 303 (1982)

Photosynthetic Potential of Plant Cell Cultures

Yasuyuki Yamada
Research Center for Cell and Tissue Culture, Faculty of Agriculture, Kyoto University, Sakyo-ku, Kyoto 606, Japan

Photoautotrophism is an outstanding characteristic of plant metabolism. Cultured green cells from higher plants now provide new potentials for research into photosynthesis, and the establishment of photoautotrophism in cultured cells is helping to advance studies on the productivity of plant cells.

Photoautotrophic cultures of various types of green cells able to grow relatively well for long periods were isolated in the late 1970s. The research which established such cultures demonstrated that the production of successful photoautotrophic cultures of green cells depends on the selection of highly chlorophyllous cells with high photosynthetic potential.

Both photoautotrophically and photomixotrophically cultured green cells mainly fix CO_2 through the Calvin cycle, but they also have a special carboxylation pathway (PEPCase) which is much more active than the usual carboxylation pathway present in intact plant cells. Very active $^{14}CO_2$-fixation into malate in the light was found for both types of cultured cells from C_3 plants; dark fixation alone could not account for this.

The study of photoautotrophism in plant cell cultures is of a fundamental nature, but the findings are important in the development of applications such as the improvement of photosynthesis, resistance to herbicides and the production of useful compounds.

1 Introduction

Haberlandt [1] was the first to attempt tissue culture. In 1902, he used cells of *Tradescantia reflexa*, and although he failed to obtain viable cultures, his work on and theory of tissue culture served as a spur to other scientists. Gautheret reported the first successful culture of cells in 1934 [2]. The color of the cultured cells was green. In 1965, Stetler and Laetsch reported that tobacco cells cultured with medium containing kinetin developed chloroplasts [3,4,5]. Other scientists followed with studies that confirmed the presence of photosynthetic activity in cultured green cells [6-14]. Light was also shown to have an important function in the growth of cultured cells. In addition, most cultured cells that are chlorophyllous have been shown to require sugar for growth [12,15].

There have been many attempts to culture chlorophyllous tissues autotrophically, with growth being maintained for short periods or at low growth rates [6-10]. Lack of vigorous autotrophic growth has been attributed partly to low chlorophyll content and low photosynthetic activity in the chlorophyllous cells, as well as to deficiencies in the culture conditions used for chloroplast development and photosynthesis.

A number of studies have been carried out on the basic physicochemical conditions and nutritional requirements for the growth and greening of chlorophyllous cells: Auxins [6,11,16], cytokinins [5,12,17], sugars [12,17,18-20], inorganic nutrients [21,22], temperature [19] and the gas phase [7,9,11,14,23,24] are all factors that have been reported to affect growth.

In the late 1970s, two groups, Hüsemann and Barz [11], and Yamada and Sato [14], independently produced photoautotrophic cultures of different types of green cells which grew relatively well for long periods. Their research demonstrated that the establishment of successful photoautotrophic cultures of green cells depends on the selection of highly chlorophyllous cells with high photosynthetic potential.

2 Selection of Photoautotrophic Cells

Cultured cells are heterogeneous in their specific characters. Therefore, cells that have the desired specific characters must undergo repeated selection in addition to being grown under carefully regulated culture conditions. Photoautotrophism is specific to only some of the cells in a green culture. Even under light, the colors of cultured cells will differ, some are white, others yellow, yellow-green or green. In the case of rice, bar-

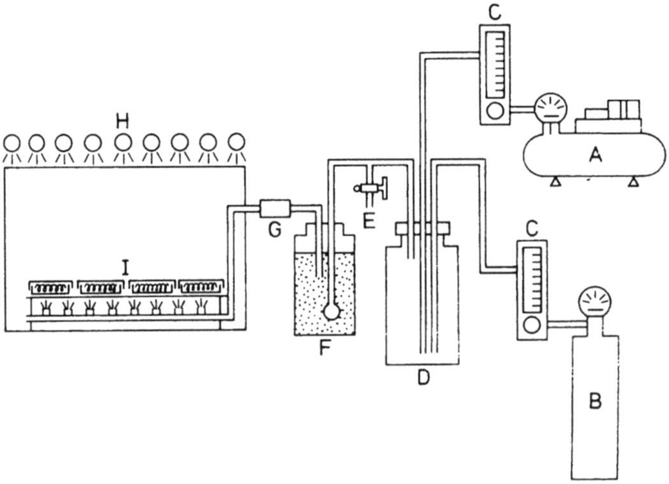

Fig. 1. An efficient system for the selection of photoautotrophic cells.
A Air compressor; B CO_2-gas; C Flow control; D Reservoir for mixed gas; E Safety valve; F Distilled water to wash the gas; G Air line filter; H Illumination; I Callus culture on medium without sugar

Table 1. Conditions used for the photoautotrophic culture of green cells

Plant cells	Auxin	Light intensity	CO_2 concentration	Refs.
Asparagus officinalis	NAA 5.4×10^{-6} M	120 μE m^{-2} s^{-1}	2% CO_2, 49% N_2	30)
Chenopodium rubrum	2,4-D 10^{-7} M	8000 lux	2%	25)
Cytisus scoparius	NAA 10^{-5} M	8000 lux	1%	14)
Datura stramonium	NAA 5×10^{-6} M	3000–5000 lux	1%	27)
Glycine max	NAA 5.4×10^{-6} M	200–300 μE m^{-2} s^{-1}	5%	29)
Hyoscyamus niger	NAA 10^{-5} M	3000–5000 lux	1%	27)
Marchantia polymorpha	2,4-D 4.5×10^{-6} M	photon flux density of ca. 90 μmol m^{-2} s^{-1} (400–700 nm)	1%	28)
Nicotiana tabacum				
'JWB su'	NAA 1.6×10^{-6} M	120–235 μE m^{-2} s^{-1} (400–700 nm)	1%	7)
'Samsun'	NAA 5.4×10^{-7} M	5000–10000 lux	1%	6)
	NAA 10^{-5} M	8000 lux	1%	14)
'Wis 38'	2,4-D 5×10^{-7} M	2500 lux	2%	8)
Ruta graveolens	NAA 1.1×10^{-6} M	2000 lux	1% CO_2, 99% N_2	9)
Spinacia oleracea	NAA 5.4×10^{-7} M	41 mE h^{-1} per fermenter	a mixture of air and N_2, plus 1% CO_2	31)

ley and oat cultures, all the cells are white or yellow-white. Green spots seen on the white calluses are the primordial germ cells of green shoots, not undifferentiated green cells.

The heterogeneity of chlorophyllous cells has led to attempts to select green cells that have photosynthetic activity. Hüsemann and Barz [11,25] continued selective subculturing of *Chenopodium rubrum* cells that had survived under photoautotrophic conditions until they obtained cell suspension with high growth rates. Yamada and Sato [14,26] selected highly chlorophyllous cells of tobacco and Scotch broom during mixotrophic culture and succeeded in photoautotrophically culturing those cells by repeated selection. These studies show that cultured cells do differ in their photosynthetic potentials and that the selection of cells with high photosynthetic potential is essential if successful photoautotrophic cultures are to be established.

Recently, an efficient method for selecting photoautotrophic cells from cultured heterogeneous cells has been established by Yasuda et al. [27]. Segments of excised leaves from redifferentiated, aseptically grown seedlings are inoculated on sugar-free nutrient agar medium in petri dishes. The hormones used in combination in the medium are 5 or 10 μM NAA and 0.05 or 5 μM BA. The inoculated petri dishes are placed in 20-liter transparent glass cabinets, and, from the beginning of callus induction, the cultures in these cabinets are aerated with a mixture of 1% CO_2 in air at a flow rate of 1 liter per minute. Illumination from fluorescent lamps is continuous at an intensity between 5000 and 8000 lux. The temperature inside the cabinets is maintained between 27 and 29 °C. In subsequent subcultures made after callus induction, the cells are cultured under these same photoautotrophic conditions, which include an adequate intensity of illumination and aeration with CO_2-enriched air. It is difficult, however, to culture any type of green cell photoautotrophically under aeration with ordinary air (ca. 0.03% CO_2).

To date, no successful cultures of green cells of cereals have been reported. Callus induction from cereals requires a high concentration of auxin in the medium, and this inhibits chloroplast differentiation.

3 Photosynthetic Potential of Cultured Green Cells

Many factors affect the photosynthetic activities of cultured green cells. Studies of the development of chloroplasts in cultured green cells have shown that the lamellar structure of the chloroplast partially disappears during the early stages of culture, then redevelops during the logarithmic phase of growth. Those chloroplasts with a developed lamellar structure clearly show CO_2 fixation and photosynthetic activity. Photosynthetic activities of cultured green cells, some of which are taking up 100–200 μmol mg^{-1} h^{-1} of $^{14}CO_2$ (refering to chlorophyll) similar to those of green leaves, are shown in Table 2. The early products of cultured photoautotrophic C_3 plant cells during CO_2 fixation differ greatly from those of C_3 plant leaves. The products of short term $^{14}CO_2$ fixation in light have been reported for photomixotrophically cultured cells obtained from C_3 plants [10,3,40,42,12], C_4 plants [43,44,45] and CAM plants [46]. Nato et al. [41] reported that the $^{14}CO_2$ fixation pathway in light differs with the growth phase. During the exponential phase of growth, the pathway of $^{14}CO_2$ fixation linked to PEPCase was greatly enhanced [42]. In cultured cells from

Table 2. Photosynthetic potential of green cultured cells

Plant cells	Chlorophyll contents	Photosynthetic activity (CO_2 fixation or oxygen evolution*)		Refs.
	μg g^{-1} fresh weight (or mg g^{-1} dry weight**)	μmol mg^{-1} h^{-1} chlorophyll	μmol g^{-1} h^{-1} fresh weight (or mg g^{-1} h^{-1} dry weight***)	
Arachis hypogaea	130–210	120–210	15–43	38)
Asparagus officinalis	2.5** (photoauto.)	230*	580*, ***	30)
	3.7** (photomixo.)	160*		37)
Chamaecereus sylvestrii	105–120	130–150 (100–115*)		35)
Chenopodium rubrum	190–200	180–190	36–38	36)
Cytisus scoparius	135	36*	4.8	26)
Datura stramonium	130	26	3.4	27)
Daucus carota	68	7.5	0.5	18)
	73	116 (Hill reaction)		12)
Gisekia pharnaceoides	150	76 (69*)	12 (10*)	38)
Glycine max	270–450	17–92	20*	29)
Hyoscyamus niger	31	140	4.3	27)
Kalanchoe fedtschenkoi	3.2**	180 (140*)	580 (450*)***	39)
Marchantia polymorpha	10–23**	2–11 (1–33*)	30–130 (16–373*)***	27, 34)
Nicotiana tabacum				
'Samasun'	4.2**	160		6)
	61	88*	5.4*	26)
'Wis. 38'	1.3**	360*		8)
'Xanthi'	1.0**	90*		41)
	1.3**	23–57*	30–80***	32, 33)
Portulaca oleracea	90–160	70–120	6–20	38)
	120	11	1.4*	40)
Spinacia oleracea	1.1–1.4**	410–550*	440–640*, ***	31)
Streptanthus tortuosus	100	12	1.2	40)
Green leaves	500–2000	100–200		
	10–50**			

C_4 plants [43,45], the four carbon acids were the most heavily labelled products of $^{14}CO_2$ fixation in light over a short period. Green callus cultures of *Kalanchoe crenata* (a CAM plant) showed no crassulacean acid metabolism, but contained greater quantities of malate, citrate and isocitrate than did colorless callus [46].

Nishida et al. [47] studied the distribution of $^{14}CO_2$ among the C_3 compounds linked to the C_1—C_5 carboxylation catalyzed by RuBPCase and the C_4 compounds linked to the C_1—C_3 carboxylation catalyzed by PEPCase in photoautotrophically-cultured tobacco cells (C_3 plant) under light [47]. As a result of short term $^{14}CO_2$ fixation under light, malate (C_1—C_3 carboxylation) was heavily labelled, as were phosphoglyceric acid and sugar phosphates (C_1—C_5 carboxylation). Two carboxylation pathways linked to RuBPCase and PEPCase were indicated in photoautotrophically- and photomixotrophically-cultured C_3 plant cells [47,48]. Results of an enzymological study supported this observation, and indicated that the carboxylation reactions are associated with high PEP carboxylase activity, the key enzyme for C_1—C_3 carboxylation [49].

The RubBP carboxylase activity in cultured cell extracts is about 1/10 that of leaf extracts of C_3 or C_4 plants on a chlorophyll basis. In contrast, the PEP carboxylase activity of cultured green cells was about 4- to 8- fold higher than that of leaf extracts of C_3 plants. on a chlorophyll basis (Table 4). Thus, when compared to C_3 leaf tissue, cultured green cells show a large increase in the PEP carboxylase/RuBP carboxylase

Table 3. Labelling patterns of C_3 and C_4 compounds from $^{14}CO_2$ fixation in light in photomixotrophic and photoautotrophic cells [47]

Time (s)	10	30	60	120	300
	(cpm × 10^5 per g fresh wt.)				
Photomixotropic cells					
Total ethanol soluble fraction	1.69	8.08	15.2	34.3	66.2
	(100)	(100)	(100)	(100)	(100)
C_3 compounds	0.76	4.52	9.1	20.2	39.7
	(45)	(56)	(60)	(59)	(60)
C_4 compounds	0.93	3.56	6.1	14.1	26.5
	(55)	(44)	(40)	(41)	(40)
Photoautotrophic cells					
Total ethanol soluble fraction	1.48	2.27	7.14	13.5	56.5
	(100)	(100)	(100)	(100)	(100)
C_3 compounds	0.92	1.36	5.36	10.0	42.9
	(62)	(60)	(75)	(74)	(76)
C_4 compounds	0.56	0.96	1.78	3.5	13.6
	(38)	(40)	(25)	(26)	(24)

Labelled compounds were grouped according to the origin of their carbon chains.
C_3 compounds are linked to the Calvin cycle and the related metabolic pathways, the glycolate pathway and glycolysis. C_4 compounds are linked to C_1—C_3 carboxylation and its related metabolic pathway, the TCA cycle.
Figures in parentheses are the percentages of ^{14}C from the total ethanol soluble fraction in each compound

Table 4. Enzyme activities related to CO_2 fixation in green cultured cells[49] RuBPCase, RuBP carboxylase;.PEPCase, PEP carboxylase; PEPCK, PEP carboxykinase. N. D., not detected

Cells	RuBPCase	PEPCase refered to chlorophyll mol mg^{-1} h^{-1}	PEPCK
N. tabacum			
Photomixotrophic	36	130	N. D.
Photoautotrophic	39	177	N. D.
C. scoparius			
Photomixotropic	37	79	N. D.
Photoautotrophic	37	79	N. D.

ratio. This is consistent with the higher percentage of labelling of C_4 acids in cultured cells than is normally found in C_3 leaf tissue.

Interestingly, C_3 plants show a high incorporation of $^{14}CO_2$ into C_4 acids at certain physiological stages. In tobacco, high CO_2 fixation and enzyme activity for the C_4 dicarboxylic acid cycle takes place in the upper, younger leaf[50]. In wheat and oats, tissues surrounding the grain exhibit relatively high CO_2 fixation by PEP carboxylase [51]. Barz proposed that the function of PEP carboxylase is in anapleurotic CO_2 fixation[52].

Recently, Sato et al. presented the results of pulse-chase feeding experiments with [U-^{14}C]-malate, which showed that carbon from malate is not donated to the Calvin cycle[49]. Furthermore, $\delta^{13}C$ measurements clearly have shown that RuBP carboxylase is the major enzyme in photosynthesis in photoautotrophic cells[53]. At present, the role of high PEP carboxylase activity in photoautotrophically- and photomixotrophically-cultured C_3 plant cells cannot be explained.

4 Photoautotrophic Jar-Fermenter Culture

Cultured green tobacco cells grown photoautotrophically under a stream of air enriched with 1% CO_2 had very low carbonic anhydrase activity (0–9% of that in intact leaves)[54]. The CO_2 compensation point for photosynthesis in photoautotrophically cultured green tobacco cells was in excess of 0.3 mM $NaHCO_3$ at pH 7.8, but that of cells isolated from tobacco leaves was less than 0.1 mM $NaHCO_3$ at pH 7.8[54]. In consequence, these cultured green cells cannot grow photoautotrophically under ordinary air because of the high CO_2 compensation point in photosynthesis. This high CO_2 compensation point reflects the optimum aeration, agitation and light intensity levels for required photoautotrophic cell culture. In a 5 liter (working volume) photoautotrophic culture aerated with 1% CO_2-enriched air (one aeration volume per medium volume per min (vvm)), agitated with a marine-type impeller (200 rpm) and illuminated at an intensity of 8000 lux, the mass of the tobacco cells did not increase because stimulated cell respiration compensated for photosynthetic activity[55]. A low oxygen supply was essential for photoautotrophic culture in the jar-fermenter. The fresh weight of the green tobacco cells increased 2-fold under photoautotrophy after 17 days of culture with an agitation speed of 200 rpm, aeration of 0.8 vvm, air containing 1% CO_2, 14% O_2 and 85% N_2 and an illumination of 8000 lux (Figs. 2, 3).

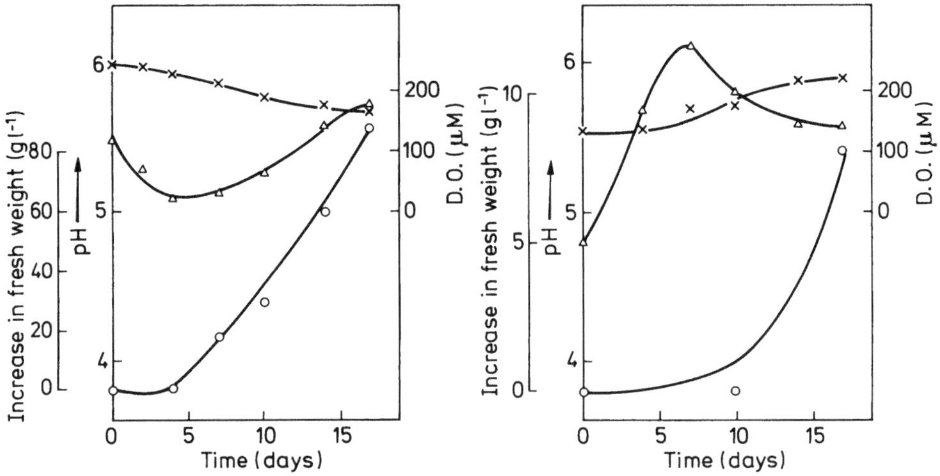

Fig. 2. Growth kinetics of green tobacco cells cultured photomixotrophically in a jar-fermenter. About 50 g of green tobacco cells (fresh weight) was inoculated in 5 l of medium containg 3% sucrose. Cells were cultured at an agitation speed of 200 rpm, an aeration rate of 1 vvm, and an illumination of 8,000 lux at 26 C. —○—, Increase in fresh weight; —△—, pH; —×—, Dissolved oxygen concentration (D.O.)[55]

Fig. 3. Growth kinetics of green tobacco cells cultured photoautotrophically in a jar-fermenter. About 50 g of green tobacco cells (fresh weight) was inoculated in 5 l of medium without sucrose. Cells were cultured at 26 °C with an agitation speed of 200 rpm, an aeration rate of 0.8 vvm using air with a lowered partial pressure of oxygen (O_2 14%, N_2 85%), an enrichment of 1% CO_2 and an illumination of 8,000 lux. Symbols are the same as in Fig. 2[55]

Hüsemann reported that the photoautotrophic growth of *Chenopodium rubrum* in a 1.5 liter airlift fermenter was 270% (cell number) after 14 days at a flow rate of 200 ml per min of 2% (v/v) CO_2-enriched air under 8000 lux light intensity[56]. Airlift systems seem to be more suitable than impeller systems for large scale culture allowing sufficient aeration and avoiding shear stress. The regulation and balance of respiratory and photosynthetic activities is very important in large scale photoautotrophic cultures.

5 Prospective Uses of Photoautotrophic Cells

The structure of chloroplasts in photoautotrophic cells and in mesophyll cells in the green leaf appear similar but the products of photosynthesis differ. Increase in PEP carboxylase activity and enhancement of malate accumulation (C_1—C_3 carboxylation), a primary product of $^{14}CO_2$ fixation, are characteristics of CO_2 fixation in the light for cultured, green C_3 plant cells. The similarities and differences between photoautotrophic cells and mesophyll cells can be used to study the mechanism of photosynthesis and to better define the nature of cultured green cells.

The photoautotrophic cell does not require any sugar in the medium since it can use CO_2 as the sole carbon source. This means that not only can carbohydrates be omitted from the medium but that green cells will produce their own characteristic carbon compound(s). Carbon metabolism is essential to all cells and the nature of the carbon source (sugar or CO_2 fixation) supplied may affect secondary metabolism and the production of useful compounds. Also, it has recently been reported that the chloroplast-encoded membrane polypeptide (32000 dalton) in chloroplasts is linked in some way to resistance to the herbicide, atrazine [57]. If we could select specific cells that are resistant to a specific herbicide from cultured green cells, we could use them as the source of a useful gene for resistance to that herbicide. The nature of the study of photoautotrophism in plant cell culture is a fundamental one, but the findings from this basic research are being developed into important applications.

6 References

1. Haberlandt, G.: Sitz.-Ber. Mat.-Nat. Kl. Kais. Akad. Wiss. Wien *111*, 69 (1902)
2. Gautheret, R. J.: C. R. Acad. Sci. *198*, 2195 (1934)
3. Laetsch, W. M., Stetler, D. A.: Amer. J. Bot. *52*, 798 (1965)
4. Stetler, D. A., Laetsch, W. M.: ibid. *51*, 667 (1964)
5. Stetler, D. A., Laetsch, W. M.: Science *149*, 1387 (1965)
6. Bergman, L.: Planta *74*, 243 (1967)
7. Berlyn, M. B., Zelitch, I.: Plant Physiol. *56*, 752 (1975)
8. Chandler, M. T., et al.: Can. J. Bot. *50*, 2265 (1972)
9. Corduan, G.: Planta *91*, 291 (1970)
10. Hanson, A. D., Edelman, J.: ibid. *102*, 11 (1972)
11. Hüsemann, W., Barz, W.: Physiol. Plant. *40*, 77 (1977)
12. Neumann, K.-H., Raafat, A.: Plant Physiol. *51*, 685 (1973)
13. Yamaya, T., et al.: Soil Sci. Plant. Nutril. *23*, 59 (1977)
14. Yamada, Y., Sato, F.: Plant & Cell Physiol. *19*, 691 (1978)
15. Bergmann, L., Balz, A.: Planta *70*, 285 (1966)
16. Sunderland, N.: Ann. Bot. *30*, 253 (1966)
17. Kaul, K., Sbharwal, P. S.: Plant Physiol. *47*, 691 (1971)
18. Edelman, J., Hanson, A. D.: Planta *98*, 150 (1971)
19. Kumar, A.: Phytomorphology *24*, 96 (1974)
20. Pamplin, E. J., Chapman, J. M.: J. Exp. Bot. *26*, 212 (1975)
21. Vasil, I. K., Hildebrandt, A. C.: Planta *68*, 69 (1966)
22. Wilmar, J. C., et al.: Nature *202*, 1235 (1964)
23. Dalton, C. C., Street, H. E.: In vitro *12*, 485 (1976)
24. Hanson, A. D., Edelman, J.: Planta *98*. 150 (1971)
25. Hüsemann, W.: Some aspects of photoautotrophic cell cultures, in: Proc. Intl. Symp. Plant Cell Culture, BPT-Report, (eds. Alferman, A. W., Reinhard, E.), p. 161, Ges. S. U. mbH., Münich 1978
26. Sato, F., et al.: Plant and Cell Physiol. *20*, 193 (1979)
27. Yasuda, T., et al.: ibid. *21*, 929 (1980)
28. Katoh, K., et al.: Planta *144*, 509 (1979)
29. Horn, M. E., et al.: Plant Physiol. *72*, 426 (1983)
30. Peel, E.: Plant Sci. Lett. *24*, 147 (1982)
31. Dalton, C. C.: J. Exp. Bot. *31*, 791 (1980)
32. Brangeon, J., Nato, A.: Physiol. Plant. *53*, 327 (1981)

33. Nato, A., et al.: ibid. *53*, 335 (1981)
34. Katoh, K.: ibid. *57*, 67 (1983)
35. Seeni, S., Gnanam, A.: ibid. *49*, 465 (1980)
36. Hüsemann, W., et al.: Protoplasma *100*, 101 (1979)
37. Seeni, S., Gnanam, A.: Plant Physiol. *70*, 815 (1982)
38. Seeni, S., Gnanam, A.: Plant & Cell Physiol. *24*, 1033 (1983)
39. Seeni, S., Gnanam, A.: ibid. *22*, 1131 (1981)
40. Kennedy, R. A.: Plant Physiol. *58*, 573 (1976)
41. Nato, A., et al.: Physiol. Plant. *41*, 116 (1977)
42. Nato, A., Mathieu, Y.: Plant Sci. Lett. *13*, 49 (1978)
43. Kennedy, R. A., Barnes, J. E.: Plant Physiol. *59*, 600 (1977)
44. Laetsch, W. M., Kortschak, H. P.: ibid. *49*, 1021 (1972)
45. Usuda, H., et al.: Plant & Cell Physiol. *20*, 193 (1979)
46. McLaren, I., Thomas, D. R.: New Physiol. *66*, 693 (1967)
47. Nishida, K., et al.: Plant & Cell Physiol. *21*, 47 (1980)
48. Hüsemann, W., et al.: Protoplasma *100*, 101 (1979)
49. Sato, F., et al.: Plant Sci. Lett. *20*, 91 (1980)
50. Kisaki, T., et al.: Plant & Cell Physiol. *14*, 505 (1973)
51. Wirth, E., et al.: Z. Pflanzenphysiol. *82*, 78 (1977)
52. Barz, W., Hüsemann, W.: Aspects of Photo Autotrophic Cell Suspension Cultures, in: Pl. Tiss. Cul. 1982 (ed. Fujiwara, A.), p. 245, Tokyo, IAPTC, 1982
53. Yamada, Y., et al.: Photosynthetic Carbon Metabolism in Cultured Photoautotrophic Cells, in: Pl. Tiss. Cul. 1982 (ed. Fujiwara, A.), p. 249, Tokyo, IAPTC, 1982
54. Tsuzuki, M., et al.: Plant & Cell Physiol. *22*, 51 (1981)
55. Yamada, Y., et al.: ibid. *22*, 919 (1981)
56. Hüsemann, W.: Protoplasma *113*, 214 (1982)
57. Hirschberg, J., et al.: Science *222*, 1346 (1983)

Selection and Screening Techniques for Plant Cell Cultures

Jochen Berlin and Florenz Sasse
GBF — Gesellschaft f. Biotechnologische Forschung m.b.H., Mascheroder Weg 1, D-3300 Braunschweig, FRG

Variants isolated from cultured plant cells are an important tool for furthering our understanding of plant metabolism and development, and they may also increase the impact of tissue cultures in biotechnology. This survey describes the techniques by which variants can be recovered from callus, suspensions or plated cells. The possibilities and the limits of analytical screening and of positive or negative selection systems are discussed.

1 Introduction

It is now generally accepted that genetic variability arises spontaneously in all types of plant cell cultures [1–5]. Indeed it has been suggested that a cell or tissue culture cycle should be included for increasing the genetic variation in plant species [3]. Larkin and Scowcroft have introduced the term "somaclonal variation" for genetic variation occurring in plants following a tissue culture cycle [3]. Some spectacular results have been reported in particular for certain polysomatic plants. For example, 15% of the regenerates of sugarcane cultures were resistant to the eyespot disease while the mother plant was susceptible [6]. Similarly, potato plants resistant to the fungal disease Early Blight were recovered by somaclonal variation [7]. The significance and problems of somaclonal variation for plant improvement have been discussed [2–5,8,9]. Despite some very interesting results one should still regard the importance of somaclonal variation for plant improvement with cautious optimism. The commercial impact will, at least partially, depend on the progress of devising suitable screening and selection schemes for the recovery of useful somaclonal variants. Our rudimentary understanding of regeneration phenomenon imposes yet a further severe hurdle. The heterogeneity of plant cell cultures provides not only the basis for the isolation of agronomically desirable mutant plants but also for the establishment of highly productive lines for natural product synthesis. The potential of plant cell cultures as a future source of natural products has recently been reviewed [10,11]. Again, the limited knowledge of the regulation of secondary metabolism in higher plants does not allow predictions regarding their future commercial viability [12].

One prerequisite for the use of plant cell cultures in biotechnology is the development of suitable screening and selection schemes for the process of interest. In this chapter we will discuss mainly the biological and technical requirements for screening and selection programmes with cultured plant cells, as this is crucial to the successful outcome of such programmes. Those readers more interested in detailed descriptions and evaluations of the different types of mutants and variants so far isolated from cultured cells are referred to recent reviews [4,8,13–17]. We assume genetic heterogeneity within a population of cultured cells as fact and will not discuss the molecular basis underlying the high variability induced by cell culture conditions. The reader will find this covered in recent reviews which discuss the nature of heritable variation occurring in plant cell cultures [8,18,19].

In theory, all cultured plant cells can be regarded as individual organisms which are omnipotent in possessing the complete genetic information for constructing the intact plant. Indeed, single cells or protoplasts of a few plant species have been regenerated into intact plants [20]. By comparing a cultured plant cell and the multicellular organism derived therefrom one realizes that selection and screening programmes on the cellular level are limited by certain severe restrictions and handicaps. One can only select and screen for traits which are already expressed by the cultured cell. However, only a small portion of the total genetic information is expressed in dedifferentiated cells. In particular for morphological alterations one cannot select at the cellular level as long as the regulatory aspects of morphogenesis are unknown and biochemical manipulation is inaccessible. On the other hand, traits selected for in cultured cells may be repressed in the regenerated plant. Thus, a successful selection at the cellular level may result in phenotypically unchanged plants. If regeneration is the step following a selection event, the value of the selected clone can only be recognized much later. This point emphasises that the peculiar characteristics of plant cell cultures should never be forgotten when devising schemes for the isolation of useful mutants. These and other features of plant cells (e.g. variable ploidy levels, density dependent growth, low growth rates) indicate that screening and selection with plant cells is undoubtedly more difficult to perform than with microbial systems. The progress of manipulating plant cells by screening and selection will therefore necessarily be much slower than it has been with microorganisms. However, the importance of higher plants for mankind and the need of further plant improvement justifies the increased effort needed to obtain a deeper understanding of plant metabolism. The availability of variant and mutant cells is a prerequisite for unraveling and altering regulatory controls in higher plants.

In this survey we will describe first the types of cultures one can use for screening and selection programmes, including the cultural requirements for the recovery of a desired variant cell. Then techniques will be discussed by which variants and mutants can be detected and isolated from the heterogeneous wild type population. Since the intention of this review is to analyze the technical state of the art of isolating variant cells we will not evaluate the importance of the various positive selection and screening events. We regard all variants stable at the cellular level as equally important for the progress in this field.

2 Definitions

Variation within a cell culture population can be due to genetic changes, epigenetic events or physiological responses. Genetic variations are defined as heritable alterations that are sexually transmitted to the progeny of plants regenerated from cultured cells. Genetic variations can be caused by changes of the nucleotide sequence or of the chromosome number and structure. Before using the term mutant for a selected trait it should be clearly shown that the alteration is transmitted meiotically or that a molecular analysis points to an altered gene product. In all other cases it is best to use the term variant. Epigenetic events include all alterations in the regulation of gene expression which are not due to mutation. Epigenetic events are normally stable, persist indefinitely throughout mitoses and are independent of the event that induced the change. In contrast to genetic alteration epigenetic traits are usually not expressed in regenerated plants and are not heritable through meiosis. Genetic and epigenetic changes have in common that the selected trait is stable on the cellular level. As we describe only the initial cellular stages of screening and selection programmes here, we are unable to distinguish between a mutant and a variant. Therefore we will generally use the term variant to describe only partially characterized stable alterations.

The heterogeneity of a cell population can also result from different responses of cells of different physiological states to the environmental conditions. The danger of selecting physiological states instead of variants is especially great when screening for compounds which are only formed during distinct periods of culture growth. In our context, we will call all screening events which are not stable on the cellular level for at least a few passages as physiological states. The problem of isolating physiological states instead of true variants will be discussed in Sect. 4 in detail.

Screening and selection are often used as interchangeable terms. However, we would like to define screening as a rather passive technique by which a great number of cell clones are analyzed for a certain trait and those showing the desired features are further cultivated and screened. Within a screening programme unchanged wild type cells are not killed but only eliminated by preferential isolation of cell clumps with the desired trait. Selection is an active process which deliberately favours only the survival of the wanted variant while wild type cells are killed.

3 Culture Systems for Screening and Selection Programmes

After defining a desirable objective to screen or select for, one has to think of a suitable culture system and of a suitable scheme for identification and isolation of the variant cells. Screening and selection for wanted traits can be done with callus or suspension cultures, with plated cells or protoplasts. Which system is the most suitable mainly depends on the purpose of the programme. For example, protoplasts of many plant species have not been regenerated to whole plants. If regeneration of a selected protoplast is required, but not possible, one would not start selection schemes with such species from protoplasts. Each culture system has its advantages and disadvantages. It is necessary and worthwhile to invest enough time in the optimization of a chosen culture system to be sure of providing optimal conditions for the recovery of variant cells. It is very important to know the abilities and the limits of the chosen culture system as selection and screening programmes can then be designed to give a positive outcome. In Fig. 1 we have assembled the culture systems which could be used for a screening and a selection programme.

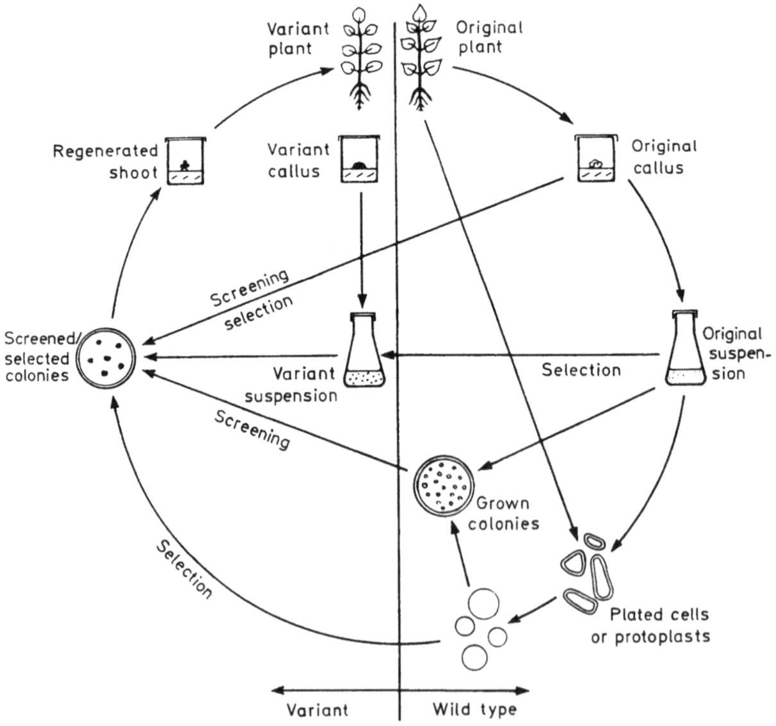

Fig. 1. The various culture systems which screening and selection programmes can be started from

3.1 Callus Cultures

The easiest, but most limited systems for a screening and selection programme are callus cultures. Consequently, the first variants of cultured cells containing increased levels of β-carotine were isolated from callus cultures [21]. Growing calluses can be checked visually for altered pigmentation by the naked eye, for fluorescent spots with a UV-lamp or for different growth behaviour. In particular screening for pigmented areas (e.g. anthocyanins [22], naphthoquinones [23], betalains [24], carotenoids [24], chlorophyll [26]) can easily and successfully be performed with calluses. Separate cultivation of areas with increased pigmentation, which sometimes has to be stabilized by repeated cloning, has provided some highly pigmented variant lines.

Visual or direct screening for altered cell clusters within a callus is a crude method of finding variants and has only a limited scope. Indeed, compared to visual screening, most other selections from calluses are rather inefficient. This is especially true with screening for resistant cells. The cells of a callus are not uniformly exposed to the culture medium and the selective agent. Thus, the recovery of a mixed population of resistant and sensitive cells is more likely than with suspended or plated cells. However, by repeatedly growing callus cells on toxic concentrations of the selective compound one can eventually eliminate these sensitive cells. Nevertheless, selection of resistant cells from calluses can only be recommended under certain circumstances e.g. when regeneration of a variant is required and is only possible from freshly initiated callus cells. If the callus cells are friable, they can often be finely dispersed on selective agar medium [27]. This minimizes the problem of escapes. Resistant colonies from calluses should be detectable after 3–8 weeks.

3.2 Suspension Cultures

The term suspension culture is used for fine suspensions of single cells and very small aggregates but also for cultures of morphological undifferentiated plant cells grown in liquid medium. Thus, a selection programme with suspension cultures can be performed with a fine, rapidly growing suspension consisting of aggregates of up to 50 cells or with a rather lumpy culture. With the latter, one will have the same problems as with callus cultures when selecting for resistance. However, fine suspension cultures seem to be particularly suitable in selections for resistance. The selective agent is added to the culture medium at a concentration which kills all sensitive cells. All cells are equally affected by the toxic compound and thus the sensitive ones should be eliminated. The disadvantage, however, is that the surviving resistant cell population, will most likely be a mixture of cells being resistant for several biochemical reasons. A freshly selected antimetabolite resistant cell population may contain cells which are resistant due to a decreased uptake of the toxic compound [28], due to the overproduction of compounds diluting the toxic effect of the agent [29,30], due to a better detoxification of the selective agent [31] or due to other biochemical alterations. A cell population consisting of a mixture of variants may be tolerable. However, the use of such variant mixtures may complicate and confuse further biochemical and genetic experiments. Consequently, single cell cloning (see below) of the selected resistant suspension is generally required, if one needs defined variants in ensuing experiments.

If one starts a selection programme for resistance in suspension cultures, one would normally detect growth of a resistant population after 4 weeks, sometimes only after 8–12 weeks. When 1.0 g fresh mass (corresponding to 2×10^7 cells) of a culture with a doubling time of 1 d is inoculated into 100 ml selective medium and only one resistant cell is expected to be present in the flask, it takes 25 days before the resistant population reaches 1 g fresh mass. Only after this time does the resistant population become visible within the mass of growth inhibited and dying sensitive cells. The number of flasks recommended for a typical selection procedure can vary greatly depending on the selection agent. We normally start with a minimum of 25 flasks. A prerequisite for a selection experiment with suspension cultures is therefore sufficient shaker capacity. Suspended cells in liquid medium should only be used for positive selection schemes but not for analytical screenings.

3.3 Plating of Cells and Protoplasts

To avoid the problems inherent in screening and selection programmes with callus and suspension cultures, the plating of cells or protoplasts on solid media or matrices seems to be an ideal alternative. However, screening and selection programmes with plated cells cause new technical problems which require overcoming. One has first to prepare single cells or small sized aggregates for plating. Plating of cells is only advantageous if the cells are plated at low densities allowing growth of individual, clearly separated colonies that can be isolated and subcultured. However, unlike microbial cells, cultured plant cells require a minimum inoculum density for growth.

3.3.1 Preparation of Single Cells and Protoplasts for Plating

The easiest way to obtain a suspension of single cells is to filter successively cells of a fine suspension culture through sieves of decreasing mesh-width from 500–75 μm. The filtered suspension will consist of single cells and small cell aggregates which are most likely of single cell origin. If the cultures are too lumpy to provide enough single cells one can try to disrupt the aggregates by adding pectinase to the culture medium 1 or 2 days before sieving. For rapidly growing rather fine suspensions immobilization of the cells in alginate has been suggested. Single cells were then released from the alginate beads [32,33].

Single cells for plating can also be prepared from differentiated tissues of intact plants. This is perhaps one of the most important arguments for screening programmes with plated cells. As stated in the introduction the tissue culture itself generates variation. Thus, if one wants to isolate variants differing from the wild type cells only at the site selected for or if one has to minimize unwanted variations, one should start screening and selection programmes from cells freshly isolated from the plant rather than from a long cultivated suspension culture. It has been shown repeatedly that extended culture periods enhanced the frequency of variant colonies [5]. The most convenient tissue of a plant, for the isolation of single cells, is the leaf. Mesophyll cells can be isolated mechanically [34–36] or enzymatically [37,38]. Division, callus formation and morphological differentiation can be induced in these cells [34,36]. However, selection and screening has only been done with mesophyll cell derived protoplasts.

In order to be sure that a selected clone is indeed derived from a single cell, selection schemes should usually be started from protoplasts of cultured cells or mesophyll cells. Since reliable cell wall degrading enzymes have been commercially available the preparation of protoplasts is routine in many laboratories. Recent reviews have summarized the newest developments in the culture of plant protoplasts [20,39]. The fact remains that protoplasts of some plant species have still to be classified as recalcitrant as they do not show cell wall generation or prolonged division and cannot be regenerated into whole plants [39].

3.3.2 Plating Techniques

The technique of plating cells on solid media was introduced by Bergmann [40] in 1960 and is described in detail by Street [41]. To ensure that growing cell colonies on a plate are of single cell origin, cell suspensions must be diluted to densities at which plant cells do not grow without special precautions. The chances of recovering colonies from plated cells at various cell densities can be determined quantitatively by the plating efficiency (PE).

$$PE = \frac{\text{number of colonies/plate}}{\text{number of viable cellular units per plate}} \times 100$$

Theoretically, a colony can be formed from every plated cellular unit. The cell density at which PE is zero is the minimum effective density. It is desirable to obtain highest possible plating efficiency at low densities. Without taking precautions plating efficiencies are generally poor at densities lower than 5000 cellular units per ml. However, for the selection of wanted mutants this cell density may already be too high as growing cellular units are not clearly separated from each other. Therefore the following special measures have been developed to achieve growth at lower densities.

The occurrence of a growth lag phase whose duration is dependent on the initial inoculum size and the fact that the cells need a minimum cell density for sustaining growth suggests that the cells have first to adapt to their medium. Cells not only absorb nutrients from the medium but also release (actively or by lysis) compounds into it. Thus, after a few days the medium of a suspension culture is altered or conditioned in a way which promotes growth of the cells. A conditioned medium can be regarded as an enriched medium which may stimulate or elevate the growth of the cells to be plated. Indeed, when conditioned medium and fresh medium are mixed in a ratio of up to 1:1 the critical initial cell density may sometimes be lowered by a factor of 10 [41]. Higher admixtures of conditioned medium cannot be recommended as the growth promoting effect may then be reversed. For example, phosphate is taken up completely from the medium by *Nicotiana tabacum* cells within 24 h [42]. Conditioned medium provided from *Nicotiana* suspensions would be devoid of phosphate. The lack of phosphate may then be the growth limiting factor.

It is evident that the use of conditioned media has in general a beneficial effect on low density growth. However, sometimes the effects are not obvious. This may be due to the fact that the composition of conditioned media is undefined and varies depending on plant species, growth rate and the time period of culturing. Thus, Kao and Michayluk [43] developed an enriched plant cell culture medium containing additional vitamins, organic acids, sugars and sugar alcohols, amino acids, nucleic acid bases, phytohormones, coconut water and casein hydrolysate. With this supplemented medium they were able to grow cells and protoplasts of *Vicia faba* at low densities of 1–50 cells per ml. This complex medium has not found the broad application one would have expected from the impressive results achieved with *Vicia* cells. Caboche [44] failed to define a completely synthetic medium able to support growth of tobacco mesophyll protoplasts at low initial densities. During his intensive studies on the nutritional requirements of tobacco cells grown at low cell densities he noted that high concentrations of naphthaleneacetic acid (NAA) were required for maximal protoplast division. On the other hand these concentrations were toxic for protoplasts cultured at low densities. As he was unable to find a concentration of NAA which promoted division without being toxic, he used an alternative approach. He precultured the cells at high initial densities in the presence of high concentrations of NAA for 4 days. The protoplasts had conditioned the medium and were then able to sustain growth when diluted to densities as low as 1–4 cells per ml. The inability of protoplasts to produce colonies at low densities in supplemented synthetic media was explained by their inability to sufficiently decrease the level of free NAA in the medium [44]. This method was also sucessfully applied to other *Nicotiana* species and *Petunia* [45]. The approach of preculturing a cell population at high cell densities in liquid medium for a short time before plating cells at low cell densities should be tested with other plant species, too. Whether this method is generally applicable for the selection of resistant mutants has to be shown.

Thus conditioned media or supplemented media improve the plating efficiency to a certain extent. However, the extent of improvement varies from culture to culture and from laboratory to laboratory. Observations in this field can only be regarded as a help and act as guidelines for improving the growth conditions of one's own culture system. The latter statement is also true for the various nurse and feeder techniques recommended by their respective authors. The history and progress in cultivating single cells in the presence of nurse callus has been described in detail by Street [41]. The nurse effect was first shown by Muir et al. [46] when they placed a single cell on filter paper which was lying on a piece of callus. After some weeks the single cell had grown into callus due to the nurse effect of the callus below the filter paper. This experiment provided the basis for all further improvements of nurse and feeder techniques.

Galun and co-workers were the first to plate cells in the presence of feeder layers. By X-ray irradiation they were able to prevent cell division without killing the feeder cells. Such inactivated but viable cells were mixed with agarized medium to serve as feeder layers for plated cells or protoplasts [47]. The feeding effect occurred also between unrelated plant species. Thus, they were able to show that carrot feeder layers can induce growth of *Nicotiana* leaf protoplasts and *Nicotiana* suspension cells at cell densities which are below the normal minimum effective cell density [48]. The growth promoting effect of a feeder layer depended on several parameter (e.g. cell density of the feeder layer, washing of the X-ray irradiated cells, preculture of plated cells or protoplasts, age of the feeder layer [48]). The value of the feeder layer for promoting growth at low densities is indisputable. However, X-ray irradiation of cells for inactiviation of cells cannot be performed in every laboratory. This is perhaps the reason why this technique has only be applied in a few laboratories. Special care has also to be taken that the inactivation of the feeder cells is complete, especially when plated cells and feeder layers are from unrelated species.

The problem of possibly mixing feeder cells and plated cells can be circumvented by placing a matrix between the nursing and growing cells. This basic principle already shown by Muir et al. [46], has been modified for the purpose of mutant selection and rapid growth by several groups. Weber and Lark [49] placed circular stainless steel screens with bent edges in petri dishes. A polyurethane pad covered by a membrane filter was placed a top at the screen (itself 3 mm above the bottom of the dish). The petri dish was filled with a medium for the cultivation of nurse cells so that the foam support was soaked with the culture medium. The cells to be screened were plated on the membrane filter. The advantage of this system is that the membrane filter with the plated cells can easily be transferred from one feeder plate to another. Thus, even screening for resistance can be performed in the presence of living nurse cells. Using this plating system, soybean cells resistant to azaguanine were isolated in 7 days. The feeder cells are grown in liquid medium and this raises the risk of infections during handling and transport. The double filter paper technique of Horsch and Jones [50] circumvents this by embedding the feeder cells in an agar medium which is covered by a filter paper (guard disc) on which a second smaller filter paper (transfer disc) is placed. The cells plated on this second filter paper can then be transferred completely or as isolated colonies to other feeder layer plates or to other media. Even cell growth can be monitored in a non-destructive way by weighing the filter paper with the growing colonies [51]. Another simple feeder layer technique has been suggested by Shneyour et al. [52]. They placed a cellophane membrane between feeder cells and plated cells. The transparent membrane allowed an easy follow-up of feeder and nursed cells. Conner and Meredith [53] simplified the system of Weber and Lark [49] by using a polyurethane support saturated with liquid medium on which a filter paper is placed for plating cells. Since plant cells can be immobilized onto polyurethane pads [54] the polyurethane support can also be used as feeder layer. Growth rates on filter paper were higher with polyurethane supports than with agar. The plated cells can be grown on polyurethane at extreme pH values where gelling properties of agar are inhibited. There are several reports that agar can have negative effects on the growth of plated cells. Alginate [55] and especially agarose [56] are attractive alternatives for solidification of culture media.

Undoubtedly, after some adaption all these systems are suitable for reducing the minimum effective density required for growth of plated cells of many plant species and will thereby ease the recovery of positive selection events.

3.4 Microculture of Single Cells

Although this techniques does not seem feasible for the initial stages of a selection scheme we would like to mention the possibility of cultivating single cells in microdroplet culture. The problem of the minimum effective density is here overcome by culturing the cell in a tiny volume of medium. The minimum effective density of 5000 cells per ml corresponds to single cell growth in a volume of 0.5 µl. This is the volume that a single cell can condition for growth. Using a combination of both approaches, microdroplets and conditioned medium, individual tobacco cells have successfully been grown in microchambers [57] and depressions of Cuprac dishes [58]. The smallest culture volume reported so far were microdroplets of 0.25 to 0.5 µl [59]. Gleba used a supplemented medium for this. By preculturing the protoplasts for 1–3 days in suspensions at high densities the plating efficiency was significantly increased. Further reductions of the volume can only be achieved by taking special precautions. Koop et al. [60] reduced the volume of the microdroplets to 10–25 nl. To prevent immediate evaporation of the tiny microvolumes the droplets were covered with mineral oil. These were the first reported attempts to culture individual cells in fully synthetic and unconditioned, but buffered, media. Division and callus formation occurred in 30% of the microdroplet cultures.

Ellis [61] was interested in establishing cell cultures from single cells of which the yield of secondary metabolites had been determined before by microspectrophotometry. He plated a diluted suspension of sieved cells on a supplemented medium [43], cut out agar pieces containing one single cell and determined the content of phenolics by microspectrophotometry. The thin agar block with the characterized cell was then placed on a filter paper lying on a feeder plate. Cell growth and product formation in daughter cells were subsequently measured. All the microculture techniques. involved are far from routine and are too laborious for the broader application in selection and screening programmes.

3.5 Replica Technique

It would be very helpful for screening and selection programmes with plant cells to have a reliable replica technique as used for bacteria [62]. Schulte and Zenk [63] tried to develop a replica technique suitable for plant cells. They placed a nylon net with a mesh-width of 500 µm on the master agar plate after the cells had been grown for 10–14 days. The colonies grew evently through the net. After about 20 days the net with the colonies was transferred to another petri dish and the net was removed. For cells of *Morinda citrifolia* the authors claimed an 80% transfer of colonies from the master plate to the copy plate. This apparently simple technique, however, has not found a broader application by other laboratories because the results were rather disappointing and not reproducible with other culture systems. The removal of the net from the master or the replica plate and the transfer to a new replica plate caused large scale destruction of the colonies of most cultures. Thus, a reliable replica technique for plant cells is not as yet available.

3.6 Mutagenic Treatment

Though the cell culture technique itself causes variation it may be desirable to enhance the variability by additional physical or chemical mutagenic treatments. Colijn et al. [64] reported that N-methyl-N-nitro-N-nitrosoguanide (MNNG) at 5–40 ng per ml increased significantly the frequency of DL-6-fluorotryptophan or

$HgCl_2$ resistance in *Petunia* cells. At the above concentrations the mutagenic treatment had no killing effect on *Petunia* cells. Weber and Lark [65] measured the ability of different mutagens to induce inherited changes. They used ethyl methansulfonate (EMS), methyl methansulfonate (MMS), MNNG, hycanthone and UV-light. The spontaneous frequency of rapidly growing maltose utilizing variants was 1.2×10^{-7}, the induced frequencies by the different mutagenic compounds varied from 3.6×10^{-5} (EMS) to 10^{-3} (frame shift mutagen hycanthone). These are only two of many examples showing that mutagenic treatment increases the mutant frequency in a cell population. Mutagenic treatment is often included in selection schemes, as the chances of isolating a distinct variant cell are usually 10–20 fold higher compared with untreated cells. However, the optimal conditions for mutagenesis may vary from culture to culture system. Therefore, a carefully established protocol for mutagenic treatment is required for optimal results.

3.7 Concluding Remarks

We have described general culture techniques which can be applied and modified to make a screening or selection programme more successful. The best technique for a given problem cannot be determined from the literature. As most researchers stick, understandably, to a chosen technique, comparison of the efficiencies of the different systems are not possible. However, it is evident that intensive work on the optimization of a chosen culture system to determine its capacity and its limits is a worthwhile exercise which will ease the following steps of screening and selection for variants. Isolation of variants from plated cells has only been reported for a few plant species. Typical of the situation is the fact that a recent review with the rather general title "Protoplasts and the isolation of mutants" listed only mutants selected from protoplasts of *Solanaceae* [17]. Indeed, for many plant species it has only so far been shown that viable dividing protoplasts can be prepared and then regenerated into plants [20], refined techniques of isolating mutants have often not been applied to these species. However, one can expect that the above techniques could easily be adopted to them. Isolation of variants from these species is probably only a question of time. Some other plant species, however, have already received the title "recalcitrant" [39]. Many other species are known to grow in vitro slowly and as lumpy cultures. Thus, in many cases the species itself may be the "limiting" factor for extensive screening and selection programmes. We should have it in mind that the most exciting results were achieved with rapidly growing, easily accessible plant cells. These results cannot indiscriminately be transferred to many other culture systems. Selection and screening programmes with cultured plant cells are at present only useful and possible with a limited number of plant species. Whether a desired objective is realizable depends to a great extent on the species chosen.

4 Analytical Screening

Having discussed the principal culture systems available for screening and selecting variants, we can now discuss the various methods by which distinct variants can be detected and separated. First we will describe analytical screening techniques. The common principle of all analytical screening techniques is that the cells of the wild type population are separately analyzed for the desired trait. Analytical screening is mainly used for establishing highly productive variant lines. Regeneration of the selected line is normally not required.

Some natural compounds (mostly secondary metabolites) of higher plants are of great commercial interest. As some cell cultures also synthesize and accumulate secondary metabolites they have been considered as an alternative source for the production of natural products [10-12]. Normally product yields are rather low in plant cell cultures. Thus, the development of techniques to screen for highly productive variant lines is a logical approach. Secondary metabolites are not required for the survival of cultured cells. On the contrary, in many plant cell cultures secondary metabolism is turned on when growth has ceased due to the depletion of certain nutrients [42]. This means that one can hardly establish a positive selection system for cells overproducing a special compound. The only possible way seems to be to screen directly calluses, suspensions or plated cells for increased levels of the desired compound. One can subdivide the analytical screening techniques into direct and indirect methods. Direct analytical screening includes all techniques by which the isolated clone can be directly subcultured. When the selected clone can only be evaluated after analysis of cell extracts and only a part of the clone can be subcultured, we call this indirect analytical screening.

4.1 Direct Analytical Screening

When discussing the value of callus cultures (3.1) for screening we mentioned that some secondary metabolites have an intense colour allowing direct visual detection. Altered pigmented areas of callus are easy to detect. Thus, it is not surprising that the first variant cell cultures isolated were lines with altered pigmentation. The technique of isolating altered pigmented areas from calluses of plated and nonplated cells for separated cultivation is self-evident. Rather than describing many of these variants we would like to call attention to some problems inherent in all analytical screening techniques.

Alfermann et al. [66] isolated cell clones from calluses of *Daucus carota* capable and incapable of synthesizing anthocyanins. These productive and non-productive lines maintained their characteristics over years and can therefore be regarded as variants. Individual cells of a variant line may show still different product levels. However, the mean value of accumulated product must be significantly different from that of wild type cells and must be expressed stably over a long period of time. Therefore, the isolation of differently pigmented clones does not necessarily yield new variant lines. The whole problem of establishing cell lines according to the actual yield of a secondary metabolite has been nicely demonstrated by Dougall's group [67]. They isolated clones containing low or high levels of anthocyanins from an anthocyanin producing cell culture of *Daucus carota*. After several passages the low and high producing clones were recloned by plating. The high producing clones had become heterogeneous again and yielded high and low producing subclones. The high producing subclones were also not stable and yielded a whole range of differently accumulating colonies. The decline from the high producing state increased with the number of passages and was dependent on the culture medium. In the stock culture medium all strains had similar anthocyanin levels after 4–5 passages. Data of earlier passages were not given. The low producing clones were also not stable and yielded high and low producing subclones after recloning. One can conclude that the visually selected cell clones were not truc variants but only physiologically different cells as the altered mean values were not stably maintained for prolonged periods of time. Secondary metabolism can be induced or triggered by many environmental factors. Anthocyanin formation has been shown to be dependent on phytohormone composition and phosphate levels [66,68,69]. Consequently the levels of accumulated anthocyanins are different in different media [67]. A cell suspension is a population of cells in different growth states, as it is a non-synchronous growing cell population. Therefore the environment has different effects on the individual cells of the heterogeneously growing cell suspension. The same environment may induce anthocyanin biosynthesis in a physiologically competent cell while in another cell anthocyanin biosynthesis cannot be fully expressed at the same time. Thus, the difference in anthocyanin levels of the different clones has often not a genetic basis but is due only to the different physiological competence for anthocyanin biosynthesis at a given time. We do not know how long environmentally induced physiological states are maintained in a single cell derived population. The initiation of anthocyanin biosynthesis within a cell can be seen as a change of cell differentiation. The signals influencing secondary metabolism and thus differentiation are often acting indirectly and have to be given a long time before actual changes become visible [70-72]. Regarding the physiological state as a visible part of a differentiation programme, it is likely that this state is passed on to daughter cells. Our interpretation that Dougall and co-workers [67] isolated physiologically different clones rather than variants is of course only a surmise. But the extreme instability of analytically selected cell lines often reported cannot be explained only by genetic instabilities but is due rather to our inability to distinguish, while selecting, between true variants and physiological states.

When we emphasize the fact that different product levels are more often reversible responses to the environment and not due to stable genetic or epigenetic alterations, we do not want to dismiss the approach of analytical screening. Undoubtedly, this technique has yielded interesting variant lines. That one can also isolate more stable lines from *Daucus carota* has been demonstrated by Kinnersley and Dougall [73]. The authors screened *Daucus carota* cell suspensions for 6 months (12 passages) on the basis of cell-aggregate size. The small-size class (less than 63 μm) had higher and the large-size class lower anthocyanin levels than the unscreened culture. The yield of the small-size class was increased 3-fold. But more important than the increase in yield was the observation that following the screening period the tendency to form larger aggregates was drastically reduced. Thus the authors may have selected a new variant cell line. When the cells were filtered only once through the 63 μm screen the culture returned immediately to the normal aggregate size population of the wild type culture. They assume that the small-size cultures may differ from wild type cells in the reduced availability of cytokinin. This assumption is, however, not based on the measurement of different internal cytokinin levels but on known effects of externally added cytokinin.

The most extensive screening for high and stable anthocyanin producing cultures has been done with *Euphorbia millii*. Yamamoto et al. [74] have outlined a rather simple, efficient but long lasting screening scheme. They started from *Euphorbia* calluses, divided these into many segments and placed the various segments separately on agar medium. One half of the segments were analyzed and the other half was subcultured. The reddest pieces of the plates were selected, divided and analyzed as before. They continued this screening for the reddest cell colonies for nearly 30 subculture generations. The mean value for the pigment content in the cell-aggregates became stable after 23 clonal selections. The yield was increased 7-fold compared with wild type cells. Interestingly, between the 16th and 19th subculture the maximum value increased 3-fold indicating that the greatest success during this long screening programme for a stable line was achieved relatively late. The authors state that they failed to get stable, highly pigmented strains by single cell cloning and they felt that cell-aggregate cloning is superior to single cell cloning. A similar conclusion can be drawn from the results published by another Japanese group. Yamakawa et al. [75] established grape callus cultures containing 1–1.8% anthocyanins by repeated callus screening for the reddest cells. Colonies isolated from feeder layers had up to 3.4% anthocyanins. Data about the stability of the producing strains were not given. Their experiments to obtain anthocyanin producing cell lines that are genetically homogeneous failed as almost all colonies obtained from protoplasts were white. The switching off of anthocyanin biosynthesis in almost all protoplast derived colonies at the same time cannot be explained by genetic alterations. The change from the producing to the non-producing state represents the expression of a new state of differentiation, which can be regarded as an epigenetic alteration, if the state is stably expressed, or as a transient state, if anthocyanin biosynthesis is soon resumed.

Shikonin derivatives, naphthoquinone pigments of commercial interest, are produced by callus cells of *Lithospermum erythrorhizon*. From these cultures Tabata's group established high producing strains by repeatedly subculturing only the red areas [23]. The highest producing strain yielded shikonin derivatives only on Linsmaier-Skoog agar medium but not in LS-liquid medium [76]. By adding a small amount of agar powder or agaropectin to the liquid medium the cells resumed shikonin biosynthesis [77]. When this strain was recloned on LS-medium solidified with agarose only 2% of the colonies produced shikonins. The clones producing shikonins on medium solidified with agarose were also able to synthesize shikonins in liquid LS-medium. The biochemical difference between the clones producing shikonins only in the presence of agaropectin and those clones in which pigment formation occurs without the acidic polysaccharide is not yet clear. This is the first observation that the quality of resulting variant strains can be influenced by the composition of the medium. Screening for colonies with increased capability for naphthoquinone synthesis together with the development of a production medium were so successful that a commercial production seems to be feasible [78, 79].

Direct screening does not need to be done with coloured compounds. When colourless compounds show a high fluorescence they can also be screened for by visual selection. We screened calluses of *Peganum harmala* under UV light for fluorescent spots [80]. By subculturing only these areas we were able to increase the mean content of harman alkaloids by 10-fold. However, these highly producing areas showed poor growth behaviour and lost their higher synthetic capacity soon after selection pressure was removed. Screening for fluorescent cells can also be done with the fluorescence microscope. Deus and Zenk [81] used this method to isolate colonies from *Catharanthus roseus* with high fluorescence levels which arose mainly from the accumulation of the indole alkaloid serpentine. The authors claimed to have established a strain producing nearby 400 mg ajmalicine per l. However, data concerning the stability of the strains selected by fluorescence microscopy were not given.

The most refined technique of direct analytical screening has been introduced by Ellis's group [61, 82]. He established single cell derived cultures from cells which had been analyzed for their phenolic content by microspectrophotometry. The content per cell of a compound having an absorbance maximum greater than 300 nm and a sufficient molar extinction coefficient can be monitored. Also the levels of fluorescent compounds per cell can be determined by this method. Thus, cell cultures can be started from a single, well-defined mother cell. Ellis established many true clonal lines of *Anchusa officinalis* and measured the rosmarinic acid content of randomly picked single cells of these clonal lines. Already in the early stages (10–50 daughter cells) analyses showed that many of the daughter cells had lower or higher levels than the mother cells. With further growth the clonal lines showed a distribution pattern similar to that of the stock culture. A correlation between rosmarinic acid levels of the mother cell and the culture derived therefrom was not found. From our discussion above this result is not unexpected if we assume that the actual content of each mother cell used in the experiment was determined mainly by its physiological state. Thus, the mother cells used in this experiment were evidently not different genetically. Only by repeated cloning of many lines may variant mother cells be detected in which increased productivity was genetically and not just physiologically determined.

4.2 Indirect Analytical Screening

In contrast to direct analytical screening where the screening may or may not be accompanied by analysis of cell extracts, indirect techniques necessarily require analyses of cell extracts to reveal the quality of a clone. The clone under investigation has to be divided for subcultivation and chemical analysis. Screening, which is not guided by the pigmentation of a clone, is of course a disadvantage which can normally only be overcome by testing a larger number of clones. Therefore simple techniques allowing quick analyses of many clones seem to be a necessity for indirect screening.

A rough but quick estimation of alkaloids for screening purposes is the cell squash method [83]. Ogino et al. [83] put small samples of tobacco calluses on filter paper and squeezed them thoroughly between two glass plates. The cell sap was absorbed by the filter paper which was then sprayed with Dragendorff's reagent. The alkaloid content (e.g. nicotine) of the samples was estimated semi-quantitatively from the colour reaction. Selection was started from a culture producing 0.7% nicotine/dry mass and a total of 1000 clones was analyzed by this method. After two cycles of cloning 5 high producing strains were isolated with up to 2.5% nicotine, while a third cloning did not yield any further increased nicotine levels. The lines were stable for at least 1 year during which the specific content fluctuated between 2 and 3%.

It should be clear that this technique can only be applied to cultures accumulating sufficient levels of alkaloids. If the wild type contains only traces of alkaloids this technique will not work.

A rather specific analysis of cell extracts was performed during one of the most impressive screening successes with plant cells. A group at the Japan Tobacco & Salt Public Corporation [84] has isolated strains from tobacco cell cultures producing the highest levels of ubiquinone-10 ever found in a living organism. By repeated cell cloning they established a variant line containing 5.2 mg g^{-1} dry mass which was nearly 15 times more than the ubiquinone level of the wild type population. The increase of this primary metabolite seems to be due to the selection of strains with an enhanced number and/or volume of mitochondria [85]. Analyses of the many clones had to be performed from cell extracts by a specific colour reaction or by HPLC.

Screening of large numbers of clones by the individual evaluation of their cell extracts by measuring absorption, fluorescence or colour reaction with or without previous chromatographic separation is a very time consuming and tedious enterprise. Thus, immunological method seems to be an attractive alternative. Detection and quantification of metabolites by immunological methods today represent one of the most specific and sensitive analytical techniques. Provided the test has been established the metabolite content of many small sized samples can be monitored in crude extracts. Zenk and co-workers [87] were the first to use radioimmunoassay (RIA) techniques to screen for variant lines. They developed RIA-assays for two closely related indole alkaloids, ajmalicine and serpentine, to screen for high producing cell colonies of *Catharanthus roseus* and isolated some remarkably high producing cell clones. Unfortunately, the selected lines were rather unstable and rapidly lost their capacity for high production [87]. It is not clear whether repeated cloning over a long period of time would have resulted in stable variant lines. The fact is that alkaloid formation of *Catharanthus roseus* cells depends greatly on the physiological state. Alkaloid accumulation occurs only in production media but not in growth media [86, 88, 89]. Therefore screening has to be done after the cells had been transferred to a production medium. Only cells in the correct physiological state can respond to the production medium with biosynthesis of alkaloids. Active clones cannot be maintained on the production medium but have to be transferred back to the growth medium for subcultivation. Screening for alkaloid producing variant cells of *Catharanthus roseus* involves a continuous shifting from one to the other differentiation programme. The danger of isolating metabolically different cells rather than variants seems to be even greater in this system than in systems where secondary metabolism is turned on in the growth medium. Unselected lines of *Catharanthus roseus* producing low or intermediate levels of indole alkaloids were found to have stable production characteristics [90].

Despite the remarkable results reported by Zenk's group no other reports of the RIA-technique being used for screening have been published. This is due mainly to the technical expenditure needed to carry out the tests. One has to label the target compound with tritium or iodine-125, make the target compound antigenic and prepare antibodies, and one needs a β- or γ-counter for measuring radioactivity. Therefore one should have excluded alternative solutions before deciding to establish a RIA-test for a screening programme. However, if the RIA-test has to be applied routinely over a long period of time the initial expenditure would be worthwhile as it is simple and has the possibility of making several hundred analyses per day. In typical screenings for strain improvement the RIA-technique is probably too sophisticated. The same statement is true for enzyme immunoassays although this method is technically more convenient for most laboratories. An enzyme immunoassay for glycyrrhetic acid, a commercially important plant product, has recently been established [91].

Radioimmunoassays are very sensitive as nmol or pmol levels can be determined quantitatively. Thus, when RIA-techniques were introduced into this field there was the great hope of detecting extremely rare variants. Many commercially important compounds have, as yet, not been found to accumulate in morphologically undifferentiated cells. If only very few cells of a suspension are capable of synthesizing these compounds they might be detectable by analytical tests as sensitive as the RIA-test. However, the search for such cells seems to be somewhat dubious since the reasons as to why the majority of the cells of a particular culture do not produce a compound are poorly understood. Hence a prediction as to how many cells must be screened to obtain a productive variant cannot be made. Thus, analytical screening for cell strains with increased productivity can only be successful with strains that already produce the desired compound.

4.3 Importance of Original Plant Material

There is a dispute in the literature about the importance of the original plant material for the productivity of cell cultures derived therefrom [86]. Kinnersley and Dougall [92] showed that high and low alkaloid tobacco plants, differing genetically only at the two loci for production and accumulation of nicotine, gave high and low producing cultures, respectively. This close correlation between nicotine content of calluses and plants suggests establishing cultures from genetically competent plants if one wants to screen for nicotine high yielding cell strains.

However, for most screening purposes the plant material is not genetically characterized with respect to secondary metabolism. The quantitative differences of alkaloids found in a population of plants can have a genetic basis or can result from physiological parameters. Large variations in yield were found between different cultures established from the same tobacco cultivar [93]. Plants which have the same nicotine content gave cultures with quite different alkaloid levels [93,94]. Similar observations were found with *Catharanthus* cultures [95]. Sometimes low alkaloid containing plants gave high producing cultures and vice versa. Again the basis for the different yields within the plant were unknown. Constabel et al. [96] analyzed the alkaloid spectra of 76 single cell clones derived from one leaf of *Catharanthus roseus*. Only 62% of the clones contained corynanthe-, strychnos- and aspidsperma-type alkaloids, while the others had varying alkaloid spectra. Variation of the alkaloid spectra of the cultures derived from one leaf was low when compared with spectra of cultures derived from different plants. For practical purposes it is recommended that cultures are established from different plants of a genetically competent producing variety. As long as the genetic basis for increased secondary product levels is not proven, screening at the plant level appears not to be justified.

Table 1. Examples of variant lines with increased product yields[a] established by repeated analytical screening

Plant species	Culture system	Product	Yields of		Number of screenings	Ref.
			Wild types	Variants		
Coptis japonica	Plated clones	Berberine	50 mg g^{-1} dry wt.	82 mg g^{-1} dry wt.	27	[178]
Hyoscyamus niger	Suspension culture	Hyoscyamine	20 µg g^{-1} fr. wt.	40 µg g^{-1} fr. wt.	4	[179]
Nicotiana tabacum	Plated clones	Nicotine	7 mg g^{-1} dry wt.	25 mg g^{-1} dry wt.	2	[83]
Nicotiana tabacum	Plated clones	Ubiquinone-10	0.5 mg g^{-1} dry wt.	5.2 mg g^{-1} dry wt.	13	[180]
Lavandula vera	Callus segments	Biotin	60 µg g^{-1} fr. wt.	150 µg g^{-1} fr. wt.	5	[122]
Lithospermum erythrorhizon	Callus segments	Shikonins	50 µg g^{-1} fr. wt.	1 mg g^{-1} fr. wt.	?	[23]
Euphorbia millii	Callus segments	Anthocyanins	1,1 A_{530} per mg fr. wt.	7,5 A_{530} per mg fr. wt.	23	[74]
Daucus carota	Suspension culture	Anthocyanins	1,2 A_{530} per mg dry wt.	3,5 A_{530} per mg dry wt.	8	[73]

[a]) Subclones of the variant lines may show still variation in yields. However, the mean value is significantly increased in variant lines compared to wild type cells. The maximum values may exceed the mean values greatly.

4.4 Concluding Remarks

Analytical screening seems still to be the only possible way for deliberate isolation of variants with altered secondary metabolite production characteristics. The general usefulness of this technique has been proven. Table 1 gives some examples of variant lines which were established by repeated cloning. The best results can be expected if the secondary metabolite in question is already formed in sufficient quantities in the growth medium. This technique can also be used to screen for variant lines with altered enzyme activities provided this is constitutively expressed or its kinetic behaviour during the growth cycle is known. Alfermann et al. [97] isolated strains of *Digitalis lanata* with increased capabilities for the hydroxylation of β-methyldigitoxin to β-methyldigoxin.

However, to avoid frustration one should also be aware of the limits of all analytical screens. During the early stages of a screening programme one can hardly distinguish between true variants and different physiological states. The fact that physiological states can feign variants makes it mandatory to omit all environmental changes during a screening programme which may introduce additional physiological alterations. During analytical screens the clones are picked from surroundings of growing wilde type cells. Exclusion of wild type cells during a screening programme is only avoided by single cell cloning or repeated cloning. Thus analytical screening is a time consuming and laborious technique. Indirect screening programmes usually will not exceed analyses of 2000 clones. As this is not sufficient to detect rare variants analytical screens should only be performed with productive cell cultures in order to alter the yields of a desired compound. For practical and theoretical purposes screens for rare events from seemingly non-productive cultures must fail.

5 Positive Selection Systems

Analytical screening for cell strains with altered biosynthetic potentials is a passive process and will only give positive results when the trait is sufficiently expressed in the wild type population. Rare variants can only be detected by active, positive or negative, selection systems favoring the survival of the variant cell while killing all other cells. The newly acquired characteristic of the variant cell population can have different causes which can be distinguished by its stability and heritability. If the cells are only adapted to the selective compound, then the selected characteristic is only maintained in the presence of the selective agent [98, 99]. A new trait may also be stable in the absence of the selective agent and be transmitted to daughter cells. Such traits of genetic or epigenetic origin may [100] or may not [101] be expressed in regenerated plants. In the latter case the trait may again be expressed in a callus initiated from the regenerated plant [101]. Sometimes these traits may be lost during plant regeneration [102]. In some cases it has been shown that the traits selected in cell cultures are transmitted sexually to progeny of regenerated plants as a Mendelian-factor [31, 103–107] or as a non-Mendelian cytoplasmic trait [108–111].

First we will deal with positive selections which involve all systems allowing only growth of the variant cell while all other cells are killed or at least are unable to grow.

Table 2. Examples for resistances selected in cell cultures. As culture systems were used callus (C) and suspension (S) cultures and protoplasts (P). The stability in the absence of the selective agent was measured on cellular level (I) and, when plants were regenerated (II), on plant level (R) or in secondary callus (RC) initiated from R-plants. In sexual progeny of R-plants (III) expression of the trait was measured on R or RC level. Mendelian (M, m) or non-Mendelian (N) heridity was also determined in a few cases. * = plated cells

Type of resistance	Plant species	Culture system	Characteristics			Ref.
			I	II	III	
Physical stress						
Chilling	*Nicotiana sylvestris*	S*	+	RC±	RC–	181)
UV-light	*Rosa damascena*	S	+			121)
Water stress	*Lycopersicon esculentum*	S	–			99)
Inorganic ions						
Aluminium	*Lycopersicon esculentum*	C, S*	+			182)
NaCl	*Nicotiana tabacum*	S	+	R+	R+	183, 184)
$HgCl_2$	*Petunia hybrida*	S*	+			64)
Vanadate	*Saccharum* spec.	S*	+			144)
Antibiotics						
Streptomycin	*Nicotiana tabacum*	C	+	RC+	RC+/N	108, 109, 185)
	Nicotiana sylvestris	C	+	RC+	RC+/m	186, 187)
Kanamycin	*Nicotiana sylvestris*	S	+	–		188)
Lincomycin	*Nicot. plumbaginifolia*	P	+	RC+	RC+/N	189)
Chloramphenicol	*Nicotiana sylvestris*	C	+	RC+	RC+	190)
Nucleic Acid Bases Analogues						
8-Azaguanine	*Happlopappus gracilis*	S*	+			191)
5-Bromodesoxyuridine	*Nicotiana tabacum*	C	+	RC+	RC+/M	104, 192)
5-Fluorouracil	*Daucus carota*	S*, P	+	RC+		102)
Amino acids						
L-Threonine	*Nicotiana tabacum*	S	+			193)
L-Valine	*Nicotiana tabacum*	P	+	R+	R+/M	194, 195, 196)

L-Ornithine	*Nicotiana tabacum*	S	+			119)
L-Phenylalanine	*Nicotiana tabacum*	S	+			119)
L-Lysine/L-Threonine	*Zea mays*	C	+	RC+	R+/RC+/M	197, 198)
Amino acid analogues						
4-Methyltryptophan	*Catharanthus roseus*	S	+			137)
5-Methyltryptophan	*Datura innoxia*	S*	+	R+		100)
p-Fluorophenylalanine	*Daucus carota*	S	+			117)
m-Fluorophenylalanine	*Nicotiana tabacum*	S	+			119)
Methionine sulfoximine	*Nicotiana tabacum*	C/P	+	R+	R+/M	103)
Ethionine	*Nicotiana tabacum*	S	+			116)
Selenomethionine	*Nicotiana tabacum*	S	+			199)
Amino ethylcysteine	*Oryza sativa*	C	+	+		200)
δ-Hydroxylysine	*Nicotiana tabacum*	S	+			115)
Hydroxyproline	*Daucus carota*	S	+			115)
Azatidine-2-carboxylic acid	*Daucus carota*	S*	+			201)
Selenocysteine	*Daucus carota*	S*	+			202)
Trifluoroleucine	*Valeriana walichii*	S	+			203)
Glycine hydroxamate	*Nicotiana tabacum*	S*	+	RC+		204)
L-Canavanine	*Saccharum* spec.	S*	+			144)
Herbicides						
2,4-Dichlorophenoxyacetic acid	*Citrus sinensis*	C	+			205)
Naphthaleneacetic acid	*Nicotiana tabacum*	P	+	RC+(shoots)		206)
Paraquat	*Lycopersicon esculentum*	S*	+	R±	RC+/M	207)
Asulam	*Apium graveolens*	S*	+			208)
Picloram	*Nicotiana tabacum*	S*	+	RC+	R+/RC+/M	106)
Amitrole	*Nicotiana tabacum*	C	+			209)
Isopropyl-N-phenylcarbamate	*Nicotiana tabacum*	P	+	R−		210)
Glyphosate	*Corydalis sempervirens*	S				143)
Miscellaneous compounds						
Aminopterin	*Datura innoxia*/Zea mays	C/S*	+	+		114, 211)
Methotrexate	*Nicotiana tabacum*	S*	+	R−/RC+	RC+	101)
Carboxin	*Nicotiana tabacum*	S*	+	RC+		212)
Colchicine	*Acer pseudoplatanus*	S*	−			213)

Table 2 (continued)

Type of resistance	Plant species	Culture system	Characteristics			Ref.
			I	II	III	
Abscisic acid	*Nicotiana sylvestris*	S*	+			[214]
Isonicotinic acid hydrazide	*Nicotiana tabacum*	S*	+	R+	R+	[107, 215]
Pimelic acid	*Lavandula vera*	S*				[122]
α-Amanitin	*Daucus carota*	S*	+	RC+		[216]
Acetohydroxamate	*Nicotiana tabacum*	S				[148]
Hydroxyurea	*Nicotiana tabacum*	C*	+	RC+	RC+/M	[27]
Pathogens						
Pseudomonas tabaci	*Nicotiana tabacum*	C/P	+	R+		[103]
Helminthosporium maydis	Zea mays	C	+	R+	R+/N	[110, 217]
Phytophtora infestans	*Solanum tuberosum*	C	+	R+		[218, 219]
Fusarium oxysporum	*Medicago sativa*	C	+	R+		[220]
Tobacco mosaic virus	*Nicotiana sylvestris*	C		R+	R+	[221]

Most cell lines established by positive selections are resistant lines. In Table 2 we give examples of the various types of resistances selected for, how they were established and how stable the traits were. We do not want to analyze how far the selected variants fulfilled the original intentions of the selection programme. Sometimes it was only intended to use the variants as markers in genetic experiments (e.g. fusion experiment [112]). Selections for resistances to salts, herbicides or phytotoxins were initiated for "improvement" of plants (Table 2). Some types of selection have yielded variants suitable for biochemical and genetic studies [113,114], while others yielded cell lines with increased levels of primary [29,115,116] and secondary [117-121] metabolites. However, the overall impression from the literature survey is that the selection of so many resistant cell lines is merely an indicator of the lack of more important selection schemes. Many agriculturally important characteristics cannot be selected in cell culture. Also, positive selection for cells overproducing a particular secondary metabolite or accumulating a protein of higher nutritional value are hard to think of. In addition, the situation for selecting important variants is complicated by the unfavorable ploidy levels of cultured cells. With this in mind research in this field should proceed in two directions. First, one should intensify research on establishing true, and hopefully stable, haploid lines for selection and screening programmes. Secondly, one should think of new or more defined selection schemes. The following examples show that a better understanding of the mode of action of selective agents may help to improve the chances of selecting a greater variety of distinct mutants.

Many plant cell cultures synthesize biotin from the precursors pimelic acid and alanine. However, high concentrations of pimelic acid are rather toxic to cultured cells while high concentrations of biotin are tolerated by the cells. This allowed Watanabe et al. [122] to select for pimelic acid resistant cell lines of *Lavandula vera* in order to find biotin overproducing strains. The overproducing strains metabolized or detoxified pimelic acid more rapidly to the nontoxic end product biotin, although it is not clear which of the 5 enzymes involved in the biosynthetic pathway from pimelic acid to biotin has been altered. Not all pimelic acid added was recovered as biotin. Whether a turnover of biotin, accumulation of intermediates or other detoxification reactions are responsible for this is not yet known. Further biochemical characterization may help to reveal the regulatory controls of biotin biosynthesis in higher plants.

Many resistant cell lines have been selected against antimetabolites, which are mainly analogues of primary metabolites. However, only a few of them have been characterized with regard to the biochemical alterations that made the cell resistant. Resistance to 5-methyltryptophan (5-MT) could be accounted for by an altered feedback sensitivity of anthranilate synthetase [29,123-125]. This allowed the resistant cells to overproduce L-tryptophan thus diluting the toxic effect of 5-MT. Overproduction of the corresponding natural amino acid also seems to be the biochemical explanation for resistances to other amino acid analogues [115,116].

Para-fluorophenylalanine (PFP) is regarded as an analogue of L-phenylalanine. However, growth inhibition by PFP can often only be partially reversed by phenylalanine or phenylalanine/tyrosine. This indicates that PFP may exert other toxic effects which are not related to the phenylalanine pathway. Thus PFP was found to block mitosis [126]. A direct effect of PFP on tubulin was assumed [127]. For *Aspergillus* it has been estimated that mutations at as many as 28 different loci may cause resistance to PFP [128]. In higher plants only a few of the possible target sites are known [31, 129]. The idea of using PFP as a chemical for haploidization, however, has been abandoned [8].

The mode of action of PFP on plant cells seems to be rather complex and is not fully understood. Even the resistance of the best characterized PFP-resistant cell line cannot be explained completely [31]. This tobacco cell line TX4 was originally selected by Palmer and Widholm [117]. The decreased uptake of all L-amino acids contributes partially to the overall resistance [130] as well as the higher capacity for detoxifying L-PFP to p-fluorocinnamic acid conjugates [31]. The latter is due to a 10-fold increase of phenylalanine ammonia lyase (PAL) activity [119]. This higher PAL activity of the resistant cells may also explain the 6–10-fold increased levels of cinnamoyl putrescines compared to wild type cells [119]. Only marginal enzymological alterations in the shikimic acid pathway have been found [131]. However, karyological and morphological differences between TX4 and wild type cells TX1 [132] clearly indicate that the selection for PFP-resistance resulted in a cell line with some characteristics unrelated to phenylalanine metabolism. Together with the failure to reverse growth inhibition of PFP by phenylalanine it shows that PFP is unlikely to provide variants whose resistance are easy to explain. The TX4 cells were selected in suspension. Single cell cloning, to separate the different mechanism probably involved in the overall resistance, are underway.

From our studies on PFP-resistant cell lines we realized that one can select directly for cell strains having a better potential for degrading toxic compounds. It is more and more likely that enzymes diverting primary metabolites into secondary pathways play a decisive role in the regulatory control of secondary metabolism. Only if such enzymes are turned on or induced may higher levels of secondary metabolites accumulate. The next enzymes of the pathway may be co-induced or be constitutively present in the cells. Therefore, we tried to select for cell lines showing high activity of these enzymes. Tryptophan decarboxylase (TDC) activity was found to be generally low in cell lines of *Peganum* and *Catharanthus* producing low levels of indole alkaloids. Analytically selected variant strains with higher levels of indole alkaloids had much higher TDC activity [133]. Also the transfer of nonproducing cells into a production medium increased TDC activity [89, 134]. Our selection programme for cell lines with increased TDC activity was started by checking for the most suitable toxic tryptophan analogues. Fourteen of the 15 tested tryptophan analogues were regarded as metabolic analogues as their growth inhibitory effect was reversed by L-tryptophan [135]. The toxicity of the analogues was related to their inhibitory effect on anthranilate synthetase. Thus, one would assume from the literature [29, 123–125, 136] that selection would mainly give cell lines with an altered feedback control of anthranilate synthetase. Only four of the analogues were good substrates for TDC. We used 4-methyltryptophan as the most likely selective agent for cell lines with a high capacity for detoxifying the analogue to the corresponding non-toxic amine [137]. All resistant cell lines of *Catharanthus roseus* had significantly higher TDC activity and consequently higher levels of tryptamine [137]. The better detoxification of the toxic analogues, however, cannot completely account for the overall resistance of these cell lines. From the mode of action one would also expect to find cell lines with an altered feedback sensitivity of anthranilate synthetase. However, as in the case of 5-methyltryptophan resistant *Catharanthus* cells isolated previously [138], 4-methyltryptophan resistant strains showed the same feedback inhibition pattern as wild type cells. Nevertheless, the alteration should be related to tryptophan biosynthesis or metabolism as the growth inhibitory effect of the analogue can be completely reversed by tryptophan [135]. We believe our direct selection systems for cell lines with high levels of TDC is an enrichment for cells already producing indole alkaloids in the growth medium [138]. Thus analytical screening for indole alkaloid producing cell colonies is now following the selection scheme. The central aspect of our selections, however, is that an improved knowledge of the mode of action of a toxic compound (e.g. antimetabolites) may help in the use of selection for resistance as a tool for establishing deliberately cell lines with increased levels of enzymes or specific products.

Glyphosate, a widely used non-selective, broad spectrum herbicide, is an inhibitor of 5-enolpyruvyl shikimic acid-3-phosphate (EPSP) synthase [139]. This enzyme might be the main target of herbicide action. However, the possibility that glyphosate may have different prime targets or even multiple ones in different plants cannot be excluded [140]. Treatment of cultured cells with glyphosate resulted in drastically increased levels of shikimic acid which is consistent with the idea of EPSP synthase as prime target [131, 141, 142]. Amrhein et al. [143] adapted cell cultures of *Corydalis* to grow in the presence of 10 mM glyphosate. Levels of EPSP synthase and shikimic acid were increased 30- and 100-fold, respectively. The authors were cautious and refused to call their tolerant line a variant [143]. King and Maretzki [144] isolated stable glyphosate resistant sugarcane lines but did not analyze for biochemical differences between resistant and sensitive strains. Selection and characterization of glyphosate resistant cell lines from plants, which are suggested to have different targets of herbicide action, may give definite answers to its mode of action [140, 143, 145, 146]. It would also be of great interest to see whether a resistant line with increased EPSP synthase activity will retain the high enzyme activity in regenerated plants and give glyphosate resistant plants. On the cellular level glyphosate resistant variants or adapted cells are an excellent source for characterizing EPSP synthase.

There are a few other examples where selection for resistance can be a suitable tool in establishing cell lines with distinctly enhanced enzyme levels. Thus, acetohydroxamate can be used to select for high levels of urease [147] and aminopterin for increased dihydrofolate reductase activity [30, 114].

As long as selection programmes have to be carried out mainly with diploid, aneuploid or polyploid cells, we have to realize that many biochemical variants can only be selected via resistance. To make the best of this unfavourable situation, the mode of action of the selective agent should be characterized as thoroughly as possible. Then, as has been shown, more interesting biochemical variants can be deliberately selected from plant cell cultures.

Finally, we want to mention the possibility of selecting positively for "utilizing" variants. Tobacco cell lines, able to grow on urea as sole nitrogen source, have been established due to slow adaptive increases in urease levels [147, 148]. Chaleff and Parsons [149] isolated tobacco cells which were capable of utilizing glycerol as sole carbon source. The biochemical alteration allowing the cells to grow on glycerol, as well as on several sugars, is not known [149]. Limberg et al. [150] recovered variants of soybean which were able to grow on maltose as sole carbon source. These utilizing variants may be interesting for biochemical or genetic studies. However, it should be clear that plant cells cannot compete with microorganisms which can be adapted to grow on many "wild" energy sources. The selection of cell lines growing on rather cheap non-refined carbon sources is likely of commercial interest. However, no cell lines capable of growing on molasses, malt extracts and milk whey have been reported.

6 Nutritional or Auxotrophic Mutants

Auxotrophic mutants have been an invaluable tool in the elucidation of biochemical pathways and their regulation. Auxotrophic mutants are nutritional or deficiency mutants as they cannot grow on minimal medium due to a block in a biosynthetic pathway. Only if the lack of essential biosynthetic intermediates caused by the block is overcome by supplementing the medium with suitable compounds would auxotrophs grow on minimal medium. Therefore it is clear that auxotrophs can only be selected for primary biosynthetic pathways which are absolutely necessary for survival. It is also evident that screening for auxotrophs should best be done with true haploids.

Table 3. Auxotrophic cell lines isolated by different techniques

Selection technique	Plant species	Deficiency or nutritional requirement	Ref.
Chlorate resistance	*Nicotiana tabacum*	nitrate reductase	152, 155, 156, 160)
	Datura innoxia	nitrate reductase	222)
	Rosa damascena	nitrate reductase	153)
	Nic. plumbaginifolia	nitrate reductase	154, 157)
Non-selective screening	*Datura innoxia*	pantothenate, adenine	164, 223)
	Hyoscyamus muticus	nitrate reductase, his, trp, nicotinamide	163, 224)
	Nic. plumbaginifolia	ile, leu, uracil	225, 226)
Enrichment/negative selection BUdR-technique	*Nicotiana tabacum*	hypoxanthine, biotin, p-aminobenzoic acid, arg, lys, pro	151)
	Hyoscyamus muticus	his	170)
	Nic. plumbaginifolia	ile, ile + val, his, leu, trp	171)
Arsenate selection	*Datura innoxia*	ile + val, thr, nitrate reductase	173)

As only a few haploid lines are available for such experiments the number of auxotrophic mutants of higher plants is still rather low (Table 3). Carlson [151] was the first to screen for auxotrophic mutants with cultured plant cells. However, as haploid cells of alloploid *Nicotiana tabacum* were used, he isolated only leaky auxotrophs showing poor growth behaviour on minimal medium which was stimulated by various supplements.

6.1 Chlorate Resistant Mutants

Several techniques have been applied to recover auxotrophs. A unique system, as it does not necessarily require haploid cells, can be used for the selection of nitrate reductase deficient mutants. By selecting for chlorate resistant cells many nitrate reductase minus lines have been established [152-157]. The nitrate reductase deficient cell lines cannot reduce the nitrate analogue chlorate into toxic chlorite. For some of the lines biochemical and genetic analyses of these conditionally lethal mutants have been performed [158-161].

6.2 Non-selective Total Isolation System

Normally auxotrophic mutants must be isolated either by non-selective procedures or by schemes that preferentially kill wild type cells during a reversible growth inhibition of auxotrophic cells. The non-selective, total isolation system was first described by Beadle and Tatum [162] for the isolation of auxotrophic cells of *Neurospora crassa*. This technique has recently been applied to cultured plant cells. Gebhardt et al. [163] prepared protoplasts from mesophyll cells of haploid leaves of *Hyoscyamus muticus*, mutagenized the freshly isolated protoplasts and plated the cells on a minimal medium supplemented with amino acids, vitamins and nucleic acid bases. After 3–4 weeks single cell colonies of 1 mm diameter were picked out with fine forceps and transferred to the minimal agar medium without the supplements. Colonies showing no or poor growth behaviour on the minimal medium were then retested for growth on supplemented medium before they were further analyzed for growth at different temperatures and on partially supplemented media. From a total of 29,000 cell colonies they recovered 12 variant clones. Three were auxotrophic for histidine, one clone required tryptophan and three clones nicotinamide. Two clones needed several amino acids and were chlorate resistant. One needed leucine and was nitrate reductase deficient. Also two temperature sensitive mutants were isolated. A biochemical analysis of the mutant blocks has partially been achieved. The rather laborious technique of isolating auxotrophs by non-selective culture conditions has also been successfully applied by Savage et al. [164] who recovered a pantothenate auxotroph from a haploid cell suspension of *Datura innoxia*. Also auxotrophs of some lower plants, liverworts and mosses, have been isolated by this technique [165, 166].

6.3 Isolation by Enrichment/Selection Methods

As the recovery of true auxotrophs from cultured haploid cells has now unequivocally been shown, one should screen for auxotrophs with enrichment or selective methods. Puck and Kao [167] introduced the 5-bromodesoxyuridine(BUdR)/light selection method for the isolation of auxotrophic mammalian cells. This technique can also be successfully used with cultured plant cells. The trick is that BUdR is incorporated into the DNA of growing cells which are subsequently killed by exposure to light. The time required for light exposure can be reduced with mammalian cells by treating the culture with the DNA-binding dye Hoechst 33258 2–3 h before light exposure [168].

This greatly sensitized the BUdR substituted DNA to photolysis. As vital staining of plant protoplasts with this dye is possible [169], this step is likely to work with plant cells, too, and should be included in the selection scheme. Auxotrophs, not growing on minimal medium do not incorporate the analogue and therefore are not killed by light. Recovery of the mutants can then be achieved by plating the killed wild type cells and the auxotrophs on supplemented media. By this technique Carlson [151] isolated the leaky auxotrophs mentioned above. Shimamoto and King [170] also used this technique and isolated a histidine requiring auxotroph. However, in subsequent experiments inconsistent effects of BUdR on cell growth were observed as several steps of the isolation procedure could not be standardized. Insufficient starvation time as well as insufficient photolysis may account for the somewhat irreproducible results. Negrutiu [171] isolated some amino acid requiring mutants by the BUdR-method but also complained of the uncertain mode of action of BUdR in plant cells. With the availability of auxotrophic mutants one should, however, be able to optimize and standardize the BUdR technique in reconstruction experiments. Perhaps the variable results can also be minimized by using a suitable feeder technique [47, 49, 50, 53].

Polacco [172] has suggested arsenate as a potential negative selection agent for auxotrophs in cultured plant cells. Sodium arsenate (1–2 mM) kills virtually all growing cells within 24 h. However, when growth is inhibited by the lack of reduced nitrogen or other substances required for growth, cells can survive the arsenate treatment. Polacco selected by this technique an amino acid requiring mutant from a soybean culture [172]. Horsch and King [173] confirmed the idea of using arsenate for the recovery of auxotrophs when they isolated a valine and isoleucine requiring mutant of *Datura innoxia*. They also encountered difficulties in routinely repeating the arsenate selection procedure as thousands of colonies survived arsenate treatment in some experiments while in others all cells were killed. They realized that the problem of density growth becomes more severe during selection procedures. From their counterselection experiments they concluded that plating at high density after arsenate treatment was a major source of variability. Wild type cells and adenine requiring mutants were treated with arsenate in medium lacking adenine and then plated on agar or on feeder plates. On agar plates wild type cells did not grow and the auxotrophic mutant also showed only a very poor plating efficiency. When the same population of wild type cells was grown on feeder layers 12% survived the arsenate treatment which indicated the presence of possibly new auxotrophs. Recovery of the adenine requiring mutant was 100% on feeder plates.

In conclusion, auxotrophs of cultured plant cells can be isolated by various techniques. However, with a few exceptions haploid cells seem to be required for a successful screening for deficiency mutants. The usefulness of the negative selection systems described for the rescue of auxotrophs depends clearly on the optimization of growth conditions at low densities.

7 Prospects

We have described the technical state of the art for screening and selecting variants or mutant plant cells. The main point of progress of the last few years is the development of culture systems in which density independent growth has been observed. The chances of recovering mutants are greatly enhanced by this. However, these systems have only been applied to a few species all of which are known to provide cultures with good growth characteristics. In the future the value of the improved feeder systems must be demonstrated with other important plant species that show poor growth behaviour in vitro. Nevertheless, the overall impression is that systems are now available by which a great variety of variant or mutant lines can be established. It is indisputable that many of these lines will greatly improve our understanding of regulatory and organizational controls of plant development.

Selected biochemical variants of cell cultures will have a great impact on basic research. This alone would justify the increased effort of selecting variants of cultured cells.

Advances in basic research are required to devise better and more meaningful selection programmes for applied research. It would be rather disappointing if the outcome of all selections and screens resulted only in improvement of our basic knowledge without fulfilling the high biotechnological expectations. The importance of this area for plant improvement and natural product synthesis has been claimed so often, that everything beyond this may be regarded as shortcomings. From the above evidence this would seem to be a rather shortsighted attitude. Selection and screening techniques per se cannot bring the breakthrough for a broad application of plant cell cultures as a source of natural products. Only if secondary pathways can be deliberately expressed in cultured cells, may we have a chance of producing a broad spectrum of specific plant products in large bioreactors. The signals for switching on and off secondary pathways, however, are not known. Thus, screening programmes for lines with increased product levels will, even in the near future, be restricted to systems in which biosynthesis of secondary metabolites is spontaneously expressed or induced by simple media variations. The situation that many commercially important compounds are not formed in vitro cannot be altered by intensifying screening programmes [12].

The general usefulness of somaclonal variation for the incorporation of additional traits into existing plant varieties has still to be proven. The progress since the first discussions of using somaclonal variation for plant improvement, however, has been rapid and promising at least for a few plant species [174, 175]. The broad spectrum of somaclonal variants suggests that by devising suitable selection methods many useful variations can be recovered. However, the lack of suitable direct selection systems for many agriculturally valuable traits on the cellular level remains. Many somaclonal variants can first be analyzed on the plant level. Thus, positive selection systems for resistances of agricultural importance seem to be the only straight forward approach for plant improvement. Herbicide and phytopathogen resistant crop plants selected on the cellular level are likely to be the first examples of practical importance.

So far we have not mentioned two other areas from which plant improvement is expected. The development of new hybrids by protoplast fusion is dependent on the presence of selectable markers in the hybrids [112]. Selection of the hybrids has to be done with the methods described here. The introduction of new genetic material into plant cells by the techniques of recombinant DNA is regarded as very promising for plant improvement [176]. The progress of this area may also dependent on finding the "right" transformed plant cell as the newly acquired DNA is randomly incorporated into the plant genome [177].

Acknowledgement: We would like to thank Dr. Wray for his linguistic advice.

8 References

1. Murashige, T.: Ann. Rev. Plant Physiol. *25*, 135 (1974)
2. Skirvin, R. M.: Euphytica *27*, 241 (1978)
3. Larkin, P. J., Scowcroft, W. R.: Theor. Appl. Genet. *60*, 197 (1981)
4. Chaleff, R. S.: Science *219*, 676 (1983)
5. Larkin, P. J. et al.: In: Protoplasts 1983. Lecture Proc. 6th Intl. Protoplast Symp. Basel. (Potrykus, I. et al. eds.), p. 51, Birkhäuser-Verlag, Basel—Boston—Stuttgart 1983
6. Heinz, D. H. et al.: In: Applied and fundamental aspects of plant cell, tissue, and organ culture. (Reinert, J., Bajaj, Y. P. S. eds.), p. 3, Springer-Verlag, Berlin, Heidelberg, New York 1977
7. Matern, U., Strobel, G., Shepard, J.: Proc. Natl. Acad. Sci. USA *75*, 4935 (1978)
8. Chaleff, R. S.: Genetics of higher plants: Applications of Cell Culture. Cambridge University Press, New York 1981
9. Shepard, J. F., Bidney, D., Shahin, E.: Science *208*, 17 (1980)
10. Barz, W., Ellis, B. E.: Ber. dtsch. Bot. Ges. *94*, 1 (1981)
11. Curtin, M. E.: Biotechnol. *1*, 649 (1983)
12. Berlin, J.: Endeavour *8*, 5 (1984)
13. Tabata, M. et al.: In: Frontiers of Plant Tissue Culture 1978. Proc. 4th Intl. Cong. Plant Tissue Cell Culture. (Thorpe, T. A. ed.), p. 381, Univ. Calgary 1978
14. Maliga, P.: In: Perspectives in Plant Cell and Tissue Culture, Intl. Rev. Cytol., Suppl. 11A. (Vasil, I. K. ed.), p. 225, Academic Press, New York 1980
15. Widholm, J. M.: In: Plant Tissue Culture as a Source of Biochemicals. (Staba, E. J. ed.), p. 99, CRC-Press, Boca Raton 1980
16. Henke, R. R.: Environ. Exp. Bot. *21*, 347 (1981)
17. Bourgin, J. P.: as Ref. (5), p. 43 (1983)
18. Bayliss, M. W.: as Ref. (14), p. 113 (1980)
19. Meins, F., Jr.: Ann. Rev. Plant Physiol. *34*, 327 (1983)
20. Davey, M. R.: as Ref. (5), p. 19 (1983)
21. Eichenberger, M. E.: C. R. Acad. Sci. *145*, 239 (1951)
22. Strauss, J.: Science *128*, 537 (1958)
23. Mizukami, H., Kanoshina, M., Tabata, M.: Phytochemistry *17*, 95 (1978)
24. Constabel, F.: Naturwissenschaften *54*, 175 (1967)
25. Townsley, P. M.: Develop. Indust. Microbiol. *18*, 619 (1977)
26. Yamada, Y., Sato, F.: Plant Cell Physiol. *19*, 691 (1978)
27. Keil, R. L., Chaleff, R. S.: Mol. Gen. Genet. *192*, 218 (1983)
28. Widholm, J. M.: Plant Sci. Lett. *3*, 323 (1974)
29. Widholm, J. M.: Biochim. Biophys. Acta *261*, 52 (1972)
30. Ohyama, K.: Environ. Exp. Bot. *16*, 209 (1976)
31. Berlin, J. et al.: Planta *155*, 244 (1982)
32. Morris, P., Fowler, M. W.: Plant Cell Tissue Org. Cult. *1*, 15 (1981)
33. Morris, P., Smart, N. J., Fowler, M. W.: ibid. *2*, 207 (1983)
34. Lang, H., Kohlenbach, H. W.: In: Production of Natural Compounds by Cell Culture Methods (Alfermann, A. W., Reinhard, E. eds.), p. 274, Ges. f. Strahlen- und Umweltforsch., München 1978
35. Colman, B., Mawson, B. T.: Can. J. Bot. *57*, 1505 (1979)
36. Schwenk, F. W.: Plant Sci. Lett. *23*, 147 (1981)
37. Jensen, R. G., Francki, R. I. B., Zaitlin, M.: Plant Physiol. *48*, 9 (1971)
38. Larsen, P. O. et al.: ibid. *68*, 292 (1981)
39. Dale, P. J.: as Ref. (5), p. 31 (1983)
40. Bergmann, L.: J. Gen. Physiol. *43*, 841 (1960)

41. Street, H. E.: In: Plant Tissue and Cell Culture, (Street, H. E. ed.), p. 207, Blackwell Scientific Publications, Oxford 1977
42. Knobloch, K. H.: Plant Cell Reports *1*, 128 (1982)
43. Kao, K. N., Michayluk, M. R.: Planta *126*, 105 (1975)
44. Caboche, M.: ibid. *149*, 7 (1980)
45. Muller, J. F., Missonier, C., Caboche, M.: Physiol. Plant. *57*, 35 (1983)
46. Muir, W. H., Hildebrandt, A. C., Riker, A. J.: Am. J. Bot. *45*, 589 (1958)
47. Raveh, D., Hubermann, E., Galun, E.: In vitro *9*, 216 (1973)
48. Cella, R., Galun, E.: Plant Sci. Lett. *19*, 243 (1980)
49. Weber, G., Lark, K. G.: Theor. Appl. Genet. *55*, 81 (1979)
50. Horsch, R. B., Jones, G. E.: In vitro *16*, 103 (1980)
51. Horsch, R. B., King, J., Jones, G. E.: Can. J. Bot. *58*, 2402 (1980)
52. Shneyour, Y. et al.: Plant Sci. Lett. *33*, 293 (1984)
53. Conner, A. J., Meredith, C. P.: Plant Cell Tissue Org. Cult. *3*, 59 (1984)
54. Lindsey, K. et al.: FEBS Letters *155*, 143 (1983)
55. Adaoha Mbanasa, E. N., Roscoe, D. H.: Plant Sci. Lett. *25*, 61 (1982)
56. Lörz, H. et al.: Plant Cell Tissue Org. Cult. *2*, 217 (1983)
57. Jones, L. E. et al.: Am. J. Bot. *47*, 468 (1960)
58. Kao, K. N.: Mol. Gen. Genet. *150*, 225 (1977)
59. Gleba, Y. Y.: Naturwissenschaften *65*, 158 (1978)
60. Koop, H.-U., Weber, G., Schweiger, H.-G.: Z. Pflanzenphysiol. *112*, 21 (1983)
61. Ellis, B. E.: In: Plant Tissue Culture 1982. Proc. 5th Intl. Cong. Plant Tissue & Cell Culture, (Fujiwara, A. ed.), Maruzen Co. Tokyo 1982
62. Lederberg, I., Lederberg, E. M.: J. Bacteriol. *63*, 399 (1952)
63. Schulte, U., Zenk, M. H.: Physiol. Plant. *39*, 139 (1977)
64. Colijn, C. M., Kool, A. J., Nijkamp, H. J. J.: Theor. Appl. Genet. *55*, 101 (1979)
65. Weber, G., Lark, K. G.: Genetics *96*, 213 (1980)
66. Alfermann, A. W., Merz, D., Reinhardt, E.: Planta Med. Suppl. 70 (1975)
67. Dougall, D. K., Johnson, J. M., Whitten, G. H.: Planta *149*, 292 (1980)
68. Seitz, U., Heinzmann, U.: Planta Med. Suppl. 66 (1975)
69. Dougall, D. K., Weyrauch, K. W.: Biotech. Bioeng. *22*, 337 (1980)
70. Floss, H. G., Robbers, J. E., Heinstein, P. F.: Rec. Adv. Phytochem. *8*, 141 (1974)
71. Luckner, M.: J. Nat. Prod. *43*, 21 (1980)
72. Drew, S. W., Demain, A. L.: Ann. Rev. Microbiol. *31*, 343 (1977)
73. Kinnersley, A. M., Dougall, D. K.: Planta *149*, 200 (1980)
74. Yamamoto, Y., Mizuguchi, R., Yamada, Y.: Theor. Appl. Genet. *61*, 113 (1982)
75. Yamakawa, T. et al.: as Ref. (61), p. 273 (1982)
76. Fujita, Y. et al.: Plant Cell Reports *1*, 59 (1981)
77. Fukui, H., Yoshikawa, N., Tabata, M.: Phytochemistry *22*, 2451 (1983)
78. Fujita, Y. et al.: Plant Cell Reports *1*, 61 (1981)
79. Fujita, Y. et al.: as Ref. (61), p. 399 (1982)
80. Sasse, F., Heckenberg, U., Berlin, J.: Plant Physiol. *69*, 400 (1982)
81. Deus, B., Zenk, M. H.: Biotech. Bioeng. *24*, 1965 (1982)
82. Chaprin, N., Ellis, B. E.: Can. J. Bot. *62*, 2278 (1984)
83. Ogino, T., Hiraoka, N., Tabata, M.: Phytochemistry *17*, 1907 (1978)
84. Matsumoto, T. et al.: Agric. Biol. Chem. *44*, 967 (1980)
85. Ikeda, T. et al.: ibid. *45*, 2259 (1981)
86. Zenk, M. H. et al.: In: Plant Tissue Culture and its Biotechnological Application. (Barz, W., Reinhardt, E., Zenk, M. H. eds.), p. 27, Springer-Verlag, Berlin—Heidelberg—New York 1977
87. Zenk, M. H.: as Ref. (13), p. 1 (1978)
88. Knobloch, K. H., Berlin, J.: Z. Naturforsch. *35c*, 551 (1980)
89. Knobloch, K. H., Berlin, J.: Plant Cell Tissue Org. Cult. *2*, 333 (1983)
90. Kutney, J. P. et al.: Tetrahedron *39*, 3781 (1983)
91. Kanaoka, M. et al.: Chem. Pharm. Bull. *29*, 1533 (1981)
92. Kinnersley, A. M., Dougall, D. K.: Planta *149*, 205 (1980)
93. Kinnersley, A. M., Dougall, D. K.: ibid. *154*, 447 (1982)

94. Tabata, M., Hiraoka, N.: Physiol. Plant. *38*, 19 (1976)
95. Roller, V.: as Ref. (34), p. 95 (1978)
96. Constabel, F. et al.: Plant Cell Reports *1*, 3 (1981)
97. Alfermann, A. W. et al.: as Ref. (86), p. 125 (1977)
98. Hasegawa, P. M., Bressan, R. A., Handa, A. K.: Plant Cell Physiol. *21*, 1347 (1980)
99. Bressan, R. A., Hasegawa, P. M., Handa, A. K.: Plant Sci. Lett. *21*, 23 (1981)
100. Ranch, J. P. et al.: Plant Physiol. *71*, 136 (1983)
101. White, D. W. R.: as Ref. (61), p. 451 (1982)
102. Sung, Z. R., Jacques, S.: Planta *148*, 389 (1980)
103. Carlson, P. S.: Science *180*, 1366 (1973)
104. Marton, L., Maliga, P.: Plant Sci. Lett. *5*, 77 (1975)
105. Bourgin, J. P.: Mol. Gen. Genet. *161*, 225 (1978)
106. Chaleff, R. S., Parsons, M. F.: Proc. Natl. Acad. Sci. USA *75*, 5104 (1978)
107. Zelitch, I., Berlyn, M. B.: Plant Physiol. *69*, 198 (1982)
108. Maliga, P., Sz.-Breznovits, A., Marton, L.: Nature New Biol. *244*, 29 (1973)
109. Maliga, P., Sz.-Breznovits, A., Marton, L.: Nature *255*, 401 (1975)
110. Gengenbach, B. G., Green, C. E., Donovan, C. M.: Proc. Natl. Acad. Sci. USA *74*, 5113 (1977)
111. Umbeck, P. F., Gengenbach, B. G.: Crop Science *23*, 584 (1983)
112. Lázar, G. B.: as Ref. (5), p. 61 (1983)
113. Berlyn, M. B.: Theor. Appl. Genet. *58*, 19 (1980)
114. Shimamoto, K., Nelson, O. E.: Plant a *153*, 436 (1981)
115. Widholm, J. M.: Can. J. Bot. *54*, 1523 (1976)
116. Gonzales, R. A., Das, P. K., Widholm, J. M.: Plant Physiol. *74*, 640 (1984)
117. Palmer, J. E., Widholm, J. M.: ibid. *56*, 233 (1975)
118. Gathercole, R. W. E., Street, H. E.: New Phytol. *77*, 29 (1976)
119. Berlin, J., Kukoschke, K. G., Knobloch, K. H.: Planta med. *42*, 173 (1981)
120. Berlin, J. et al.: J. Nat. Prod. *45*, 83 (1982)
121. Murphy, T. M., Hamilton, C. M., Street, H. E.: Plant Physiol. *64*, 936 (1979)
122. Watanabe, K., Yano, S. I., Yamada, Y.: Phytochemistry *21*, 513 (1982)
123. Widholm, J. M.: Biochim. Biophys. Acta *279*, 48 (1972)
124. Carlson, J. E., Widholm, J. M.: Physiol. Plant. *44*, 251 (1978)
125. Sung, Z. R.: Planta *145*, 339 (1979)
126. Sisken, J. E.: Chromosoma *44*, 91 (1973)
127. Morris, N. R., Oakley, C. E.: J. Gen. Microbiol. *114*, 449 (1979)
128. Srivastava, S., Sinha, U.: Genet. Res. *25*, 29 (1975)
129. Lea, P. J., Norris, R. D.: Phytochemistry *15*, 585 (1976)
130. Berlin, J., Widholm, J. M.: Z. Naturforsch. *33c*, 643 (1978)
131. Berlin, J., Witte, L.: ibid. *36c*, 210 (1981)
132. Berlin, J., Sasse, F., Knobloch, K. H.: as Ref. (61), p. 329 (1982)
133. Sasse, F., Heckenberg, U., Berlin, J.: Z. Pflanzenphysiol. *105*, 315 (1982)
134. Knobloch, K. H., Hansen, B., Berlin, J.: Z. Naturforsch. *36c*, 40 (1981)
135. Sasse, F., Buchholz, M., Berlin, J.: ibid. *38c*, 910 (1983)
136. Scott, A. I., Mizukami, H., Lee, S. L.: Phytochemistry *18*, 795 (1979)
137. Sasse, F., Buchholz, M., Berlin, J.: Z. Naturforsch. *38c*, 916 (1983)
138. Schallenberg, J., Berlin, J.: ibid. *34c*, 541 (1979)
139. Steinrücken, H. C., Amrhein, N.: Biochem. Biophys. Res. Commun. *94*, 1207 (1980)
140. Rubin, J. L., Gaines, C. G., Jensen, R. A.: Plant Physiol. *70*, 833 (1982)
141. Amrhein, N. et al.: ibid. *66*, 830 (1980)
142. Holländer-Czytko, H., Amrhein, N.: Plant Sci. Lett. *29*, 89 (1983)
143. Amrhein, N. et al.: FEBS Letters *157*, 191 (1983)
144. King, J., Maretzki, A.: Physiol. Plant. *58*, 457 (1983)
145. Nafziger, E. D., Widholm, J. M., Slife, F. W.: Plant Physiol. *71*, 623 (1983)
146. Lee, T. T.: Physiol. Plant. *54*, 289 (1983)
147. Skokut, T. A., Filner, P.: Plant Physiol. *65*, 995 (1980)
148. Yamaya, T., Filner, P.: ibid. *67*, 1133 (1981)
149. Chaleff, R. S., Parsons, M. F.: Genetics *89*, 723 (1978)
150. Limberg, M., Cress, D., Lark, K. G.: Plant Physiol. *63*, 718 (1979)

151. Carlson, P. S.: Science *168*, 487 (1970)
152. Müller, A. J., Grafe, R.: Mol. Gen. Genet. *161*, 67 (1978)
153. Murphy, T. M., Imbrie, C. W.: Plant Physiol. *67*, 910 (1981)
154. Marton, L. et al.: Mol. Gen. Genet. *182*, 301 (1982)
155. Evola, S. V.: ibid. *189*, 447 (1983)
156. Buchanan, R. J., Wray, J. L.: ibid. *188*, 228 (1982)
157. Negrutiu, I., Dirks, R., Jacobs, M.: Theor. Appl. Genet. *66*, 341 (1983)
158. Mendel, R. R. et al.: Mol. Gen. Genet. *181*, 395 (1981)
159. Mendel, R. R., Alikulov, Z. A., Müller, A. J.: Plant Sci. Lett. *27*, 95 (1982)
160. Müller, A. J.: Mol. Gen. Genet. *192*, 275 (1983)
161. Evola, S. V.: ibid. *189*, 455 (1983)
162. Beadle, G. W., Tatum, E. L.: Am. J. Bot. *32*, 678 (1945)
163. Gebhardt, C., Schnebli, V., King, P. J.: Planta *53*, 81 (1981)
164. Savage, A. D., King, J., Gamborg, O. L.: Plant Sci. Lett. *16*, 367 (1977)
165. Schieder, O.: Mol. Gen. Genet. *144*, 63 (1976)
166. Ashton, N. W., Cove, D. J.: ibid. *154*, 87 (1977)
167. Puck, T., Kao, F.: Proc. Natl. Acad. Sci. USA *58*, 1227 (1967)
168. Stetten, G., Latt, S. A., Davidson, R. L.: Somat. Cell Genet. *2*, 285 (1976)
169. Meadows, M. G., Potrykus, I.: Plant Cell Reports *1*, 77 (1981)
170. Shimamoto, K., King, P. J.: Mol. Gen. Genet. *189*, 69 (1983)
171. Negrutiu, I.: as Ref. (206), p. 158 (1983)
172. Polacco, J. C.: Planta *146*, 155 (1979)
173. Horsch, R. B., King, J.: ibid. *159*, 12 (1983) and *160*, 168 (1984)
174. Evans, D. A., Sharp, W. R.: Science *221*, 949 (1983)
175. Ammirato, P. V. et al.: Trends in Biotechnol. *2*, 53 (1984)
176. Barton, A. K., Brill, W. J.: Science *219*, 671 (1983)
177. Ursic, D., Slightom, J. L., Kemp, J. D.: Mol. Gen. Genet. *190*, 494 (1983)
178. Sato, F., Yamada, Y.: Phytochemistry *23*, 281 (1984)
179. Hashimoto, T. et al.: as Ref. (61), p. 305 (1982)
180. Matsumoto, T. et al.: as Ref. (61), p. 275 (1982)
181. Dix, P. J.: Z. Pflanzenphysiol. *84*, 223 (1977)
182. Meredith, C. P.: Plant Sci. Lett. *12*, 17 (1978)
183. Nabors, M. W. et al.: ibid. *4*, 155 (1975)
184. Nabors, M. W. et al.: Z. Pflanzenphysiol. *97*, 13 (1980)
185. Menczel, L. et al.: Theor. Appl. Genet. *59*, 191 (1981)
186. Maliga, P. et al.: Mol. Gen. Genet. *172*, 13 (1979)
187. Maliga, P.: Theor. Appl. Genet. *60*, 1 (1981)
188. Dix, P. J., Joô, F., Maliga, P.: Mol. Gen. Genet. *157*, 285 (1977)
189. Cséplö, A., Maliga, P.: Current Genet. *6*, 105 (1982)
190. Dix, P. J.: Ann. Bot. *48*, 315 (1981)
191. Horsch, R. B., Jones, G. E.: Can. J. Bot. *56*, 2660 (1978)
192. Maliga, P., Marton, L., Sz.-Breznovits, A.: Plant Sci. Lett. *1*, 119 (1973)
193. Heimer, Y. M., Filner, P.: Biochim. Biophys. Acta *215*, 152 (1970)
194. Bourgin, J. P., Chupeau, M. C., Missonier, C.: In: Variability in plants regenerated from tissue culture. (Earle, E. D., Demarly, Y. eds.), p. 163, Praeger, New York 1982
195. Bourgin, J. P.: In: Genetic Engineering in Eukaryotes. (Lurquin, P. F., Kleinhofs, A. eds.), p. 195, Plenum Publ. Corp., New York 1983
196. Vunsh, R., Aviv, D., Galun, E.: Theor. Appl. Genet. *64*, 51 (1982)
197. Hibbered, K. A. et al.: Planta *148*, 183 (1980)
198. Hibbered, K. A., Green, C. E.: Proc. Natl. Acad. Sci. USA *79*, 559 (1982)
199. Flashman, S. M., Filner, P.: Plant Sci. Lett. *13*, 219 (1978)
200. Schaeffer, G. W., Sharpe, F. T., Jr.: In vitro *17*, 345 (1981)
201. Cella, R., Parisi, B., Nielsen, E.: Plant Sci. Lett. *24*, 125 (1982)
202. Furner, I. J., Sung, Z. R.: Plant Physiol. *71*, 547 (1983)
203. Becker, H., Bäumer, J. J.: Z. Pflanzenphysiol. *112*, 43 (1983)
204. Lawyer, A. L., Berlyn, M. B., Zelitch, I.: Plant Physiol. *66*, 334 (1980)
205. Spiegel-Roy, P., Kochba, J., Shoshana, S.: Z. Pflanzenphysiol. *109*, 41 (1983)

206. Muller, J. F., Caboche, M.: In: Protoplasts 1983. Poster Proc. 6th Intl. Protoplast Symp. Basel. (Potrykus, I. ed.), p. 156, Birkhäuser-Verlag, Basel—Boston—Stuttgart 1983
207. Thomas, B. R., Pratt, D.: Theor. Appl. Genet. *63*, 169 (1982)
208. Merrick, M. M. A., Collin, H. A.: Plant Sci. Lett. *20*, 291 (1981)
209. Barg, R., Umiel, N.: Z. Pflanzenphysiol. *83*, 437 (1977)
210. Aviv, D., Galun, E.: ibid. *83*, 267 (1977)
211. Mastrangelo, I. A., Smith, H. H.: Plant Sci. Lett. *10*, 171 (1977)
212. Polacco, J. C.: Ann. NY Acad. Sci. *287*, 385 (1977)
213. Zryd, J. P.: Experientia *35*, 1168 (1979)
214. Wong, J. R., Sussex, I. M.: Planta *148*, 103 (1980)
215. Berlyn, M. B.: Theor. Appl. Genet. *58*, 19 (1980)
216. Vergara, M. R. et al.: Z. Pflanzenphysiol. *107*, 313 (1982)
217. Gengenbach, B. G., Green, C. E.: Crop Science *15*, 645 (1975)
218. Behnke, M.: Theor. Appl. Genet. *55*, 69 (1979)
219. Behnke, M.: ibid. *56*, 151 (1980)
220. Hartman, C. L., McCoy, T. J., Knous, T. R.: Plant Sci. Lett. *34*, 183 (1984)
221. Murakishi, H. H., Carlson, P. S.: Plant Cell Rep. *1*, 94 (1982)
222. King, J., Khanna, V.: Plant Physiol. *66*, 632 (1980)
223. King, J., Horsch, R. B., Savage, A. D.: Planta *149*, 480 (1980)
224. Strauss, A., Bucher, F., King, P. J.: ibid. *153*, 75 (1981)
225. Sidorov, V., Menczel, L., Maliga, P.: Nature *294*, 87 (1981)
226. Sidorov, V. A., Maliga, P.: Mol. Gen. Genet. *186*, 328 (1982)

Bei Fragen zur Produktsicherheit wenden Sie sich bitte an:
If you have any questions regarding product safety,
please contact:

Walter de Gruyter GmbH
Genthiner Straße 13
10785 Berlin
productsafety@degruyterbrill.com